Island Ecology - The Need for Its Research and Education for Conservation

Edited by Eusebio Cano Carmona and Ana Cano Ortiz

Published in London, United Kingdom

Island Ecology - The Need for Its Research and Education for Conservation
http://dx.doi.org/10.5772/intechopen.1006257
Edited by Eusebio Cano Carmona and Ana Cano Ortiz

Contributors
Ana Cano Ortiz, Andrew Loerch, Douglas Stow, E. Bharathi, Edmond Hoxha, Elio Hobdari, Elva Leka, Emiliano Qose, Era Xhaferaj, Erland Alla, Eusebio Cano Carmona, Fjoralba Prifti, Genti Proko, Hilde Ervik, Hsiao-chien Shih, Jani Zoica, Jeton Pekmezi, John O'Leary, Julie Lambert, K. Priyalatha, Kellie Uyeda, Kelsey Warkentin, Kimberly O'Connor, L. Malavika, Leslie Bolick, Lloyd Coulter, Melissa L. Thomas, Neritan Ceka, Perikli Qirjazi, Shanmugam Jayalakshmi, Skënder Aliu, Spartak Kuçaj, Viktoria Smenes

First published in London, United Kingdom, 2025 by IntechOpen
IntechOpen is the global imprint of INTECHOPEN LIMITED, registered in England and Wales, registration number: 11086078, 167-169 Great Portland Street, London, W1W 5PF, United Kingdom

For EU product safety concerns: IN TECH d.o.o., Prolaz Marije Krucifikse Kozulić 3, 51000 Rijeka, Croatia, info@intechopen.com or visit our website at intechopen.com.

British Library Cataloguing-in-Publication Data
A catalogue record for this book is available from the British Library

Island Ecology - The Need for Its Research and Education for Conservation
Edited by Eusebio Cano Carmona and Ana Cano Ortiz
p. cm.

This title is part of the Environmental Sciences Book Series, Volume 31
Topic: Ecosystems and Biodiversity
Series Editor: J. Kevin Summers
Topic Editors: Salustiano Mato, Josefina Garrido and Francisco Ramil

Print ISBN 978-1-83634-659-3
Online ISBN 978-1-83634-658-6
eBook (PDF) ISBN 978-1-83634-660-9
ISSN 2754-6713

IntechOpen Book Series

Environmental Sciences

Volume 31

Aims and Scope of the Series

Scientists have long researched to understand the environment and man's place in it. The search for this knowledge grows in importance as rapid increases in population and economic development intensify humans' stresses on ecosystems. Fortunately, rapid increases in multiple scientific areas are advancing our understanding of environmental sciences. Breakthroughs in computing, molecular biology, ecology, and sustainability science are enhancing our ability to utilize environmental sciences to address real-world problems.

The four topics of this book series - Pollution; Environmental Resilience and Management; Ecosystems and Biodiversity; and Water Science - will address important areas of advancement in the environmental sciences. They will represent an excellent initial grouping of published works on these critical topics.

Meet the Series Editor

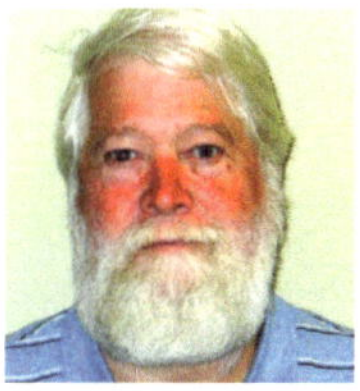

J. Kevin Summers is a Senior Research Ecologist at the Environmental Protection Agency's (EPA) Gulf Ecosystem Measurement and Modeling Division. He is currently working with colleagues in the Sustainable and Healthy Communities Program to develop an index of community resilience to natural hazards, an index of human well-being that can be linked to changes in the ecosystem, social and economic services, and a community sustainability tool for communities with populations under 40,000. He leads research efforts for indicator and indices development. Dr. Summers is a systems ecologist and began his career at the EPA in 1989 and has worked in various programs and capacities. This includes leading the National Coastal Assessment in collaboration with the Office of Water which culminated in the award-winning National Coastal Condition Report series (four volumes between 2001 and 2012), and which integrates water quality, sediment quality, habitat, and biological data to assess the ecosystem condition of the United States estuaries. He was acting National Program Director for Ecology for the EPA between 2004 and 2006. He has authored approximately 150 peer-reviewed journal articles, book chapters, and reports and has received many awards for technical accomplishments from the EPA and from outside of the agency. Dr. Summers holds a BA in Zoology and Psychology, an MA in Ecology, and Ph.D. in Systems Ecology/Biology.

Meet the Volume Editors

Eusebio Cano Carmona holds a Ph.D. in Science from the University of Granada, Spain, and is a Professor of Botany at the University of Jaén, Spain. His primary line of research is focused on flora and vegetation. His work has centered on regions including Spain, Italy, Portugal, Palestine, the Caribbean islands, and Mexico. As a result of this research, he has supervised 13 doctoral theses and produced around 265 publications, including articles, books, and book chapters. He has participated in national and international congresses with approximately 200 papers and communications. He has held various academic positions, including Director and Dean of the Faculty of Experimental Sciences. He is the founder and Director of the International Seminar on Management and Conservation of Biodiversity, which he led for 14 years. He is a member of the Spanish, Portuguese, and Italian Geobotany societies and a Counselor of the I. E. G. His academic and teaching work centers on Geobotany, teaching courses in the Biological Sciences degree program and in the master's program on olive groves and olive oil. His research has focused primarily on flora and vegetation, including the development of cartographic projects for European habitats.

Ana Cano Ortiz has a Ph.D. in Botany from the University of Jaén in Spain. She has worked in private business, in college, in secondary education and at the University of Complutense in Madrid. She is co-director of 3 doctoral theses. Her line of research is related to botanical bioindicators and the Didactics of Experimental Sciences. She has worked in Spain, Italy, Portugal and Central America. She has presented over one hundred papers, published in various national and international journals, as well as in books and book chapters, and has delivered a hundred papers or communications at national and international congresses. She is currently a professor at the Department of Didactics of Experimental, Social and Mathematical Sciences. She is a co-editor of the prestigious Journal Citation Reports (JCR) journal and a reviewer of journal articles. Her research has primarily focused on edaphic bioindicators, bioclimatology, agriculture, and science education, with applications in both European territories and Central American and Palestinian areas.

Contents

Preface

The Academic Editors would like to thank IntechOpen and Ms. Iva Horvat (Publishing Process Manager) for the opportunity to edit the book *Island Ecology – The Need for Its Research and Education for Conservation*, and, of course, all the researchers who have contributed their research to this project. This is an excellent scientific work in which the academic editors have compiled articles of high scientific value, all related to environmental aspects and methodological issues, highlighting the importance of new technologies for mapping and forecasting through aerial images of vegetation and landscapes. The authors emphasize the need to teach and understand these technologies throughout the educational system. To this end, they highlight the value of education at all levels, from basic high school to university, with the aim of training experts with environmental management skills. While all the chapters in this book are of great interest to potential readers, perhaps the lessons learned regarding the use of mapping technologies, such as GIS and drones, will draw the most attention. It also focuses on aspects of pollution and education that foster socio-environmental awareness. Environmental studies could be encouraged in schools to generate environmental awareness and advocacy.

Dr. Eusebio Cano Carmona
Professor of Botany,
Department of Animal and Plant Biology and Ecology,
Section of Botany,
University of Jaén,
Jaén, Spain

Dr. Ana Cano Ortiz
Professor of Experimental Sciences,
Faculty of Education,
Department of Didactics of Experimental Social Sciences and Mathematics,
Section of Didactics of Experimental Sciences,
Complutense University of Madrid,
Madrid, Spain

Chapter 1

Learning in Nature: Protected Spaces as Scenarios for Environmental Education

Ana Cano Ortiz and Eusebio Cano Carmona

Abstract

Natural disasters are becoming a serious global problem due to climate change and environmental pollution. This problem is causing a loss of biological diversity, social conflicts, economic crisis and damage to the health of the population. For all these reasons, we propose a new paradigm for education and research, since research and education must go hand in hand, as they are fundamental pillars for mitigating environmental problems. This study highlights a new educational model by using natural areas as teaching spaces, with very positive results in learning after practical teaching in the natural environment. A learning success rate of 100% was obtained for items 6, 7, 10, 12 and 13 related to Biosphere Reserves, while questions 1, 8, 11 and 17 show values close to 100, which are questions related to Natural and National Parks and Biosphere Reserves.

Keywords: learning, natural spaces, environmental education, climate change, holistic teaching

1. Introduction

In light of present and future environmental challenges, various researchers attribute these issues to human exploitation of the natural environment. The primary causes include climate change, habitat degradation and loss, the proliferation of invasive exotic species, as well as overexploitation and pollution [1, 2]. As a consequence, society is experiencing an increase in environmental disasters, including the disappearance of numerous ecosystem services linked to the destruction and deterioration of natural habitats.

Among these issues, the loss of biological diversity stands out as one of the most severe environmental problems facing society today, having far exceeded safety thresholds [3]. Given this reality, education must play a fundamental role in addressing environmental issues, including biodiversity loss. It is essential to integrate biodiversity and taxonomy education into teacher training programs [4]. In this regard, protected natural areas gain particular significance, as they serve a key role in conservation efforts while also providing opportunities for direct engagement with nature and on-site learning. Furthermore, these areas support educational, scientific, and cultural studies within their boundaries [5].

Considering these environmental challenges and the urgent need for solutions at the societal level, there is a pressing obligation to educate individuals in environmental values and knowledge. As Fuentealba et al. [6] highlight, behavioral and habit change at a societal level should be incorporated into environmental education objectives, both through formal and non-formal learning approaches. However, these educational efforts must be experiential in nature, ensuring that each learning experience is meaningful and leads to a comprehensive understanding of the subject matter.

A protected natural area provides an opportunity to address content from various fields of knowledge, particularly those related to living organisms, biological diversity and the environment. For future teachers to incorporate these protected areas into teaching-learning processes, it is essential to integrate their use into teacher training programs. This necessity forms the foundation of the present study.

Considering the socio-environmental crisis, such as climate change and the loss of ecosystem services, as well as the importance and necessity of mitigation efforts, and acknowledging the undeniable educational potential of protected natural areas, this study aims to address the lack of research on transmitting knowledge to future educators regarding these spaces and their application in the classroom. Thus, the study establishes the following objectives:

- To determine university students' general perceptions.
- To assess students' knowledge regarding the different existing protection categories.
- To identify students' perceptions of protected natural areas as educational resources.

1.1 Objectives

A protected natural area provides an opportunity to address content from various fields of knowledge, particularly those related to living organisms, biological diversity and the environment. For future teachers to incorporate these protected areas into teaching-learning processes, it is essential to integrate their use into teacher training programs. This necessity forms the foundation of the present study.

Considering the socio-environmental crisis, such as climate change and the loss of ecosystem services, as well as the importance and necessity of mitigation efforts, and acknowledging the undeniable educational potential of protected natural areas, this study aims to address the lack of research on transmitting knowledge to future educators regarding these spaces and their application in the classroom. Thus, the study establishes the following objectives:

- To determine university students' general perceptions.
- To assess students' knowledge regarding the different existing protection categories.
- To identify students' perceptions of protected natural areas as educational resources.

2. Methodology

1. Participants or Units of Analysis

   This research targeted university students, with a total of 128 participants, whose ages ranged from 19 to 22 years on average.

2. Variables and Instruments

   To collect data, a questionnaire specifically designed for this study was used. The questionnaire was evaluated by three experts from different universities, who suggested modifications. After incorporating their recommendations, the questionnaire was validated.

The questionnaire was administered *via* Google Forms, chosen for its user-friendly interface and ease of access through electronic devices.

The questionnaire was structured around two main dimensions:

- A fundamentals section, aimed at assessing general scientific knowledge about protected natural areas.

- A consciousness-raising section, designed to evaluate students' perceptions of protected natural areas as educational resources.

This study does not require approval from an ethics committee, as the data processing follows irreversible anonymization protocols. To ensure compliance, we adhered to the Basic Anonymization Guide published by the Spanish Data Protection Agency [7, 8] and Organic Law 3/2018, of December 5, on Personal Data Protection and Digital Rights Guarantee.

Using the 18 questions from the fundamentals section (excluding those on consciousness-raising), responses from the 128 students were compiled into an Excel spreadsheet and analyzed statistically using Past.exe. A linear correlation was calculated for each item, along with the average correlation, to apply Cronbach's alpha coefficient, which measures internal consistency and reliability [9]. The following formula was used:

$$\alpha = \frac{n \cdot p}{1 + p(n-1)} \tag{1}$$

where

- α = Cronbach's alpha coefficient

- n = Number of items

- p = Average of all correlations

2.1 Fundamentals section: Questions and response types

1. Do you know any National Parks in Spain?
    - Response type: Binary (Yes, No). (Question 1)
2. Indicate the number of National Parks in Spain.
    - Response type: Multiple-choice (108; 16; 12; 118; I do not know). (Question 2)
3. What are the functions of a Natural Park?
    - Response type: Multiple-choice (Conservation; Citizen collaboration; Research and dissemination; Leisure and tourism; Conservation, research, dissemination and tourism; I do not know). (Question 3)
4. Do you know the difference between a Natural Park and Rural Parks?
    - Response type: Binary (Yes, No). (Question 4)
5. Who is responsible for the management of National Parks?
    - Response type: Multiple-choice (Autonomous Communities; State; Private company; Others; I do not know). (Question 5)
6. Do you know what Biosphere Reserves are?
    - Response type: Binary (Yes, No). (Question 6)
7. Are you familiar with the Man and the Biosphere Programme?
    - Response type: Binary (Yes, No). (Question 7)
8. What is the function of the Man and the Biosphere Programme?
    - Response type: Multiple-choice (Conservation; Territorial development; Promoting tourism and the local economy; Conservation and sustainable development; Others; I do not know). (Question 8)
9. Are you aware of the zoning system in Biosphere Reserves?
    - Response type: Binary (Yes, No). (Question 9)
10. Are there urban centers within the territory of a Biosphere Reserve?
    - Response type: Binary (Yes, No). (Question 10)
11. How many Biosphere Reserves are there in Spain?
    - Response type: Multiple-choice (10; 103; 13; 53; None of the above; I do not know). (Question 11)

12. On the following scale, indicate your level of knowledge about the functions of Biosphere Reserves.

    - Response type: Likert scale (None; Little; Some; Quite a lot; A great deal). (Question 12)

13. Do you think citizen participation is allowed in the management of Biosphere Reserves?

    - Response type: Binary (Yes, No). (Question 13)

14. The creation of a Biosphere Reserve is proposed by...

    - Response type: Multiple-choice (Scientists; The State; The residents; The European Union; I do not know). (Question 14)

15. Do you know what Ramsar sites are?

    - Response type: Binary (Yes, No). (Question 15)

16. Are you familiar with the different environmental protection designations in Spain?

    - Response type: Likert scale (None; Few; Some; Several; Many). (Question 16)

17. Do you consider National Parks to be accessible to the general population?

    - Response type: Multiple-choice (Yes; No; I do not know). (Question 17)

18. Do you know what the Natura 2000 Network is?

    - Response type: Binary (Yes, No). (Question 18)

2.2 Awareness section: Questions and response types

1. At what level do you consider it necessary to educate about Protected Natural Areas (PNAs) in schools?

    - Response type: Likert scale (Not important; Slightly important; Somewhat important; Important; Very important). (Question 19)

2. During your secondary education, were you taught content about Protected Natural Areas and their importance?

    - Response type: Binary (Yes, No). (Question 20)

2.3 Procedure

After assessing students' prior knowledge through the questionnaire, the teaching-learning process is initiated using active methodologies, incorporating inquiry-based learning and project-based learning approaches.

During this intermediate stage, students select a specific Protected Natural Area, specifically a Biosphere Reserve. They conduct an inquiry process on the selected site and subsequently develop the foundation of their scientific project. For this phase, students utilize the CANVA platform to design a scientific poster.

Following this, a dissemination and discussion phase is carried out. Through this process, students acquire essential skills for their professional futures, such as linguistic and public speaking competencies. As emphasized by Edgar Dale [10], these educational resources should be considered pedagogical aids that enhance the teaching-learning process.

3. Results

After analyzing the initial questionnaire, as well as the assessments conducted before and after the teaching-learning phase, the following percentages of correct responses were determined (**Table 1**):

Category and function	Question number	N-initial	N-final
Fundamentals: Determining General Knowledge of Protected Natural Areas (ENP)	1	119	119
	2	40	104
	3	40	37
	4	24	43
	5	43	40
	6	30	128
	7	0	128
	8	0	110
	9	0	125
	10	21	128
	11	15	113
	12	6	128
	13	85	128
	14	0	46
	15	0	18
	16	0	70
	17	98	119
	18	27	79
Opinion: Determining Future Teachers' Personal Perception	19	125	128
	20	18	12

Note: N-initial = number of students who provide a correct answer to each question in the initial questionnaire; N-final = number of students who provide a correct answer to each question in the final questionnaire.

Table 1.
Percentage of correct responses in the initial and final objective test, categorized by question.

The linear correlation analysis r results in a value of 0.1734. When applying Cronbach's alpha coefficient, $\alpha = 18 \times 0.1734 / [1 + 0.1734(18-1)] = 0.79$. This alpha value demonstrates the internal consistency of the items used and a high degree of reliability.

The one-way ANOVA analysis follows a normal distribution pattern, as it is arranged in a line or close to it, with no outliers present (**Figure 1**). The Shapiro-Wilk value of 0.978 is close to 1, which indicates that the distribution may be normal. In the Barchart/Boxplot, the median divides the box into two asymmetric parts, with the lower part being longer than the upper part, indicating a leftward skew (**Figure 2**).

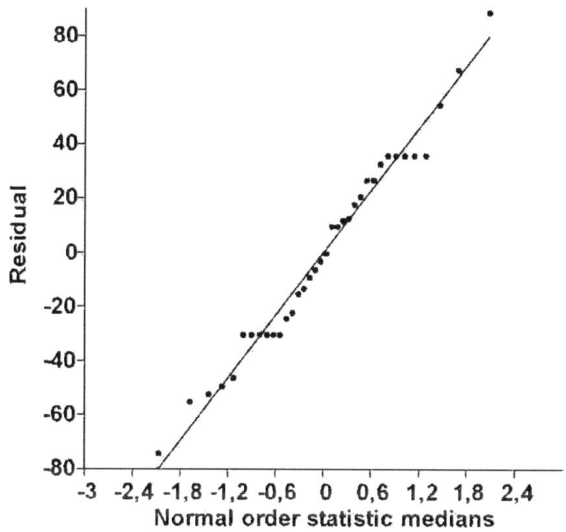

Figure 1.
ANOVA residuals with a Shapiro-Wilk value of 0.978 and p(normal) = 0.692.

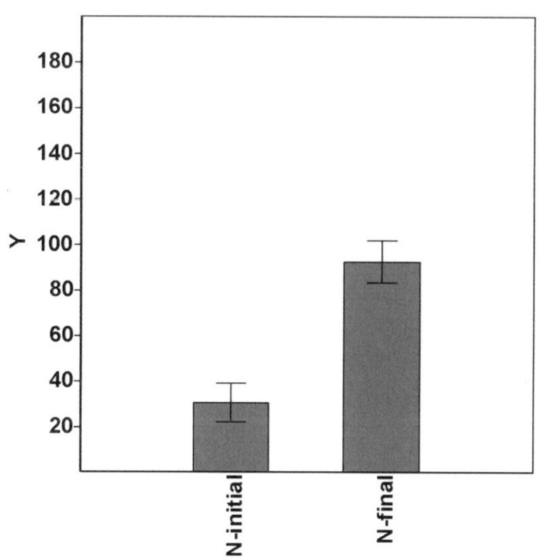

Figure 2.
Barchart/Boxplot diagram skewed to the left.

4. Discussion

As observed in **Figure 1**, from questions 6 to 13 (related to the protected area Biosphere Reserves), the greatest significant learning is located. This is because the second stage of the study, where students learn through active methodologies, focused on this natural space. This allows us to confirm that the use of this methodology (inquiry and project-based learning) is effective in the assimilation of knowledge.

Similarly, the questions grouped under the awareness block demonstrate that future educators acknowledge, both prior to and after the teaching-learning process, the importance of educating for sustainability through the use of protection figures within the school.

In this study, in addition to questions 6 to 13, questions 18 and 19 also show a very positive trend. However, there is no success in the case of questions 3, 4 and 5: *What are the functions of a Natural Park?*; *Do you know the difference between a Natural Park and Rural Parks?*; *Who is responsible for the management of National Parks?*, highlighting the students' lack of knowledge regarding natural, rural and national parks (**Figure 3**).

A detailed analysis of the items before and after learning highlights the existence of three student groups:

- *G1*: Those who mark very positively on the Likert scale with a five for most of the questions answered after the learning process (**Figure 4**), indicating teaching success.
- *G2*: Composed of questions 3, 4 and 5, where learning success is not excessive.
- *G3*: With questions 14 and 15, there is a small success, with initial negative answers of zero and positive answers after teaching 46 and 18, respectively (**Figure 5**).

The success of the learning process is attributed to the teaching method. In this case, the inquiry-based method, grounded in holistic constructivist theory, has been applied. We have followed studies related to holistic education or integral education [11–14]; Vargas and Acuña [15] incorporate constructivism as a pedagogical model of

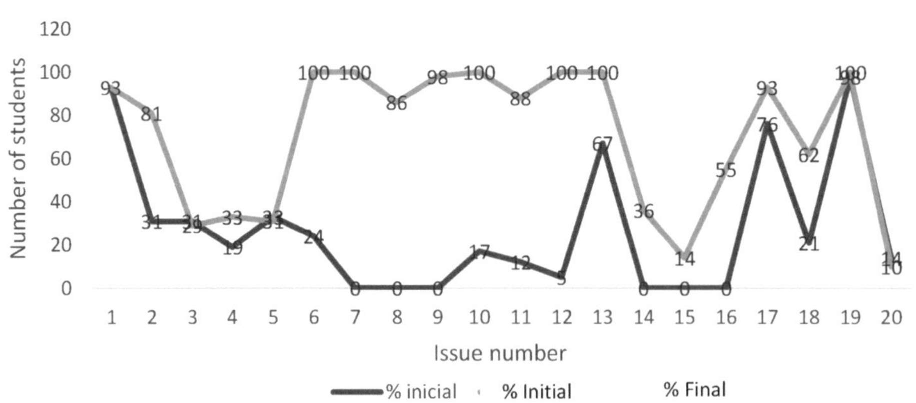

Figure 3.
Percentage of correct answers and trend comparison between the initial and final questionnaires. Note: % *initial = percentage of students who provide a correct answer to each question in the initial questionnaire;* % *final = percentage of students who provide a correct answer to each question in the final questionnaire, considering the items from the fundamental and awareness blocks.*

DOI: http://dx.doi.org/10.5772/intechopen.1010190

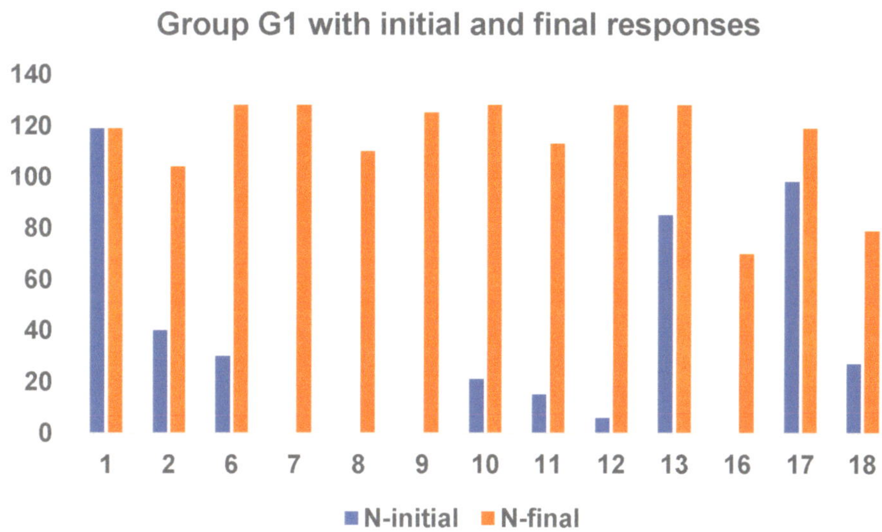

Figure 4.
Positive responses from students before and after learning (Group G1).

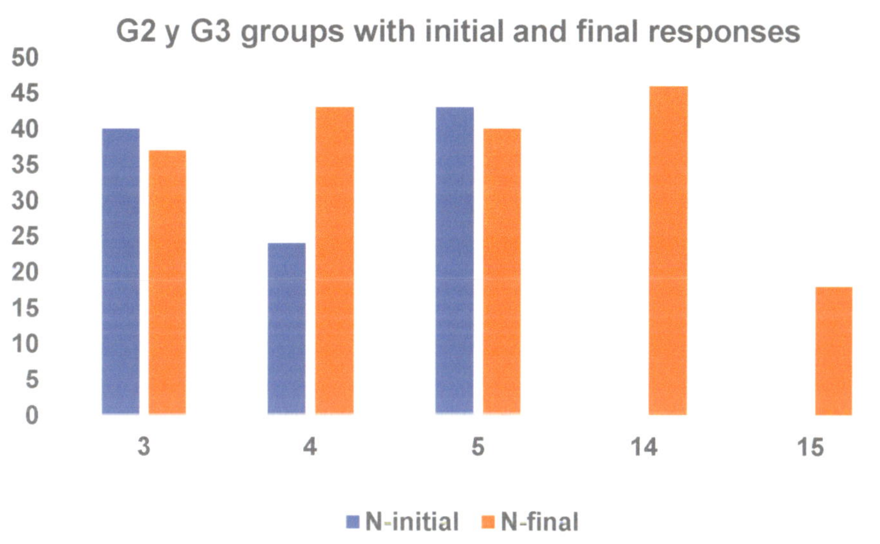

Figure 5.
Positive responses from students before and after learning (Groups G2 and G3).

interest for educators. The eminently practical inquiry-based methodology has shown to provide good training for students. Authors such as Romero Ariza, Bazán, Reyes-Cárdenas and Garritz [16–20] argue that this method strengthens the educational success of students, resulting in professional competencies regarding environmental management. In terms of educational success through active methodologies, Cano-Ortiz et al. [21] propose the use of natural and peri-urban spaces as teaching laboratories, a study in which they express the need to valorize natural and peri-urban spaces by including them in the educational system, which would involve incorporating these spaces into educational centers.

5. Conclusions

This study demonstrates that the use of active teaching methodologies results in a higher number of successes compared to traditional teaching. These methodologies prove to be innovative, through which individuals acquire knowledge and awareness of the current serious situation regarding climate change, biodiversity loss and pollution. The use of areas near cities can and should be utilized as teaching spaces. Natural Parks and Biosphere Reserves are fundamental structures for teaching about nature, as they are better-conserved areas, making them more suitable for study and interpretation by students. However, it is also observed that the peripheral areas of cities are of interest for learning due to their easy accessibility and the ability to establish comparative analysis with respect to better-conserved areas.

Conflict of interest

The authors declare no conflict of interest.

Author details

Ana Cano Ortiz[1] and Eusebio Cano Carmona[2*]

1 Department of Didactics of Experimental, Social and Mathematical Sciences, UCM, Madrid, Spain

2 Department of Animal and Plant Biology and Ecology, University of Jaen, Spain

*Address all correspondence to: ecano@ujaen.es

References

[1] Marchese C. Biodiversity hotspots: A shortcut for a more complicated concept. Global Ecology Conservation. 2015;**3**:297-309. DOI: 10.1016/j.gecco.2014.12.008

[2] Van Dyke F, Lamb RL. Conservation Biology. Foundations, Concepts, Applications. 3rd ed. Switzerland: Springer Nature; 2020. 613 p. DOI: 10.1007/978-3-030-39534-6

[3] Singh JS. Environment: A futuristic review. Current Science. 2017;**113**(2):210-217. DOI: 10.18520/cs/v113/i02/210-217

[4] Lindemann-Matthies P, Martin R, Eija YP. Professional competence of student teachers to implement species identification in schools - A case study from Germany. CEPS Journal. 2017;7(1):29-47. Available from: https://files.eric.ed.gov/fulltext/EJ1137836.pdf

[5] Crespo JM, Gómez ML, Cruz LA. Una aproximación a los Parques Nacionales y sus paisajes a través de itinerarios didácticos. Espacio, Tiempo y Forma, Serie VI Geografía. 2018;**11**:121-140. DOI: 10.5944/etfvi.11.2018.22359

[6] Fuentealba M, Marín F, Castillo F, Roco L. Análisis de la experiencia pedagógica: Campamento Explora Chile Va! Valorando la biodiversidad maulina. Actualidades Investigativas en Educación. 2017;**17**(1):1-25. DOI: 10.15517/aie.v17i1.27211

[7] Ley Orgánica 3/2018, de 5 de diciembre, de Protección de Datos Personales y garantía de los derechos digitales. BOE núm. 294, 2018. Available from: https://www.boe.es/buscar/act.php?id=BOE-A-2018-16673 [Accessed: January 14, 2025]

[8] Agencia Española Protección Datos. Guía básica de anonimización. 2022. pp. 1-71. Available from: https://www.aepd.es/documento/guia-basica-anonimizacion.pdf [Accessed: January 14, 2025]

[9] Oviedo HC, Campo-Arias A. Aproximación al uso del coeficiente alfa de Cronbach. Revista Colombiana de Psiquiatria. 2005;**34**(4):572-580. Available from: http://www.scielo.org.co/pdf/rcp/v34n4/v34n4a09.pdf

[10] Dale E. Métodos de enseñanza audiovisual. 2nd ed. México: Reverté; 1962. 573 p. Available from: https://www.iberlibro.com/M%C3%A9todos-ense%C3%B1anza-audiovisual-DALE-Edgar-Revert%C3%A9/17300975713/bd

[11] Forbes SH. In: Brandon, editor. Holistic Education: An Analysis of its Ideas and Nature. ed. Vermont, EE. UU: Foundation for Educational Renewal; 2003. 408 p

[12] Hare J. La educación holística: una interpretación para los profesores de los programas del IB. Organización del Bachillerato Internacional. Ed. International Baccalaureate; 2010. pp. 1-8. Available from: https://es.scribd.com/document/177983712/La-educacion-holistica-John-Hare [Accessed: November 8, 2024]

[13] María DS, Del holismo al constructivismo. Los grandes maestros. Revista de Postgrado FACE-UC. 2014;**6**:39-51. Available from: http://www.arje.bc.uc.edu.ve/arj15/art03.pdf

[14] Añez de Bravo Mg A. Modelo de aprendizaje holístico del ser: una propuesta pedagógica en orientación.

Revista Estilos de Aprendizaje. 2009;**3**(2):177-195. Available from: https://revistaestilosdeaprendizaje.com/article/view/884/1572

[15] Vargas K, Acuña J. El constructivismo en las concepciones pedagógicas y epistemológicas de los profesores. Revista Innova Educación. 2020;**2**(4):555-575. DOI: 10.35622/j.rie.2020.04.004

[16] Bazán LGS. La metodología indagación y el aprendizaje de las Ciencias Natura- les. Polo del Conocimiento: Revista científico – profesional. 2021;**6**(12):63. Available from: https://dialnet.unirioja.es/servlet/articulo?codigo=8219316

[17] Romero-Ariza M. El aprendizaje por indagación: ¿existen suficientes evidencias sobre sus beneficios en la enseñanza de las ciencias? Revista eureka sobre enseñanza y divulgación de las ciencias. 2017;**14**(2):286-299. DOI: 10.25267/rev_eureka_ensen_divulg_cienc.2017.v14.i2.01

[18] Reyes-Cárdenas F, Padilla K. La indagación y la enseñanza de las ciencias. Educación Química. 2012;**23**(4):415-421. DOI: 10.1016/s0187-893x(17)30129-5

[19] Garritz A. Naturaleza de la ciencia e indagación: cuestiones fundamentales para la educación científica del ciudadano. Revista Iberoamericana de Educación. 2006;**42**:127-152. DOI: 10.35362/rie420765

[20] Garritz A. Indagación: las habilidades para desarrollarla y promover el aprendiza-je. Educación Química. 2010;**21**(2):106-110. DOI: 10.1016/s0187-893x(18)30159-9

[21] Cano-Ortiz A, Piñar Fuentes JC, Rodrigues Meireles C, Cano E. Urban natural spaces as laboratories for learning and social awareness. Sustainability. 2024;**16**:3232. DOI: 10.3390/su16083232

Chapter 2

Mitigating Biosecurity Risks on Islands: Preventative Measures

Melissa L. Thomas

Abstract

Islands are biodiversity hotspots, supporting a high proportion of endemic plant and animal species, yet they face disproportionately high extinction rates due to the introduction of invasive alien species. Globalization has intensified these risks, as islands increasingly rely on shipping and air transport for supplies, and tourism for economic growth, heightening the likelihood of alien species introductions. Biosecurity management plans are established tools for preventing the introduction of invasive alien species to islands, yet only a small fraction of islands globally have such plans. A focus of these plans should be mitigating introductions through human activities, which are the primary pathways associated with invasive alien species introductions onto islands. Mitigating measures for intentional human introductions are generally easier to manage than unintentional ones, however, there remain significant social challenges associated with both pathways. A future challenge in island biosecurity will be sustaining investment in preventative measures, with international collaboration likely to play a key role in success.

Keywords: island biosecurity, alien invasive species, biosecurity management plans, pathways, mitigating strategies

1. Introduction

Despite making up less than 5% of the Earth's landmass, islands have significant global ecological importance, hosting a disproportionately high share of global biodiversity and serving as critical hotspots for species richness [1]. The isolation of island regions have made them global centers for endemic species, with islands boasting an endemism richness 9.5 times higher than that of mainland regions [2]. However, islands are also the epicenter for global extinctions, with 90% of recorded extinctions occurring on islands [3]. Extinction rates for mammals are 177 times higher in island ecosystems than the average for all ecosystems [4], while 95% of bird, and 90% of reptile extinctions have occurred on islands [5]. These extinctions are primarily the result of the introduction of alien invasive vertebrates to islands [5]. Rats, cats, and goats have played a dominant role in native species loss, devastating native populations through predation, competition, and habitat destruction [6, 7].

The expansion of global trade over the last century has intensified the problem of alien species accelerating the risk of new alien species introductions and compounding the pressures already facing islands. Many islands that were previously protected from alien species by their isolation, are now investing in modern transportation infrastructure making them more connected to global networks and the corresponding biosecurity risks (e.g. [8]). This, combined with an increasing reliance on tourism for income by island inhabitants [9], has heightened island vulnerability to biological invasions (e.g. [10]). This growing vulnerability underscores the critical importance of implementing robust biosecurity measures to protect island ecosystems. Island managers can effectively mitigate the risk associated with new alien species introductions through the development of robust biosecurity management plans.

2. Biosecurity management plans

Biosecurity management plans are proven tools for the prevention, eradication, and/or management of islands against alien invasive species [11]. However, only a small proportion of islands globally are currently governed under a biosecurity management plan [12]. There is no universal approach to island biosecurity planning. Plans differ depending on the size of the island, the location of the island relative to the mainland, the level of human activity or development on the island, and the types of ecosystems and species present. When resources are limited, biosecurity management plans will often prioritize mitigating risks associated with high-risk taxa or reducing the spread of already established invasive species to neighboring islands. For instance, biosecurity plans have been developed for the Orkney Islands, off the north-east coast of Scotland, and Guam in the western Pacific ocean, with a focus on preventing the spread of stoats [13], and the brown tree snake [14, 15], to uninvaded neighboring islands respectively.

Biosecurity plans may also be developed in response to new infrastructure that will increase visitation to the island. For example, on St Helena Island in the South Atlantic, the first airport opened in 2016 enabling private jets to arrive from Africa, Europe and South America, with commercial flights commencing in 2017 [8]. In anticipation of this air traffic, the St Helena authorities introduced a national biosecurity framework and associated policy [8]. Similarly, on the Houtman Abrolhos Islands off the coast of Western Australia, a major government initiative aimed to boost tourism by enhancing the infrastructure available for visitors on the island [16]. In response, a comprehensive biosecurity management plan was developed with a focus on mitigating the environmental risks associated with the anticipated increased human activity, while ensuring a sustainable development of tourism on the islands [17].

In the last decade there have been several guidelines, templates, and resource kits published to support effective planning and implementation of island biosecurity [9, 18, 19]. The common theme across these documents is that the most effective biosecurity plans comprise three central tenets: prevention, detection, and response. Together these measures minimize the movement of alien species to islands and maximize early detection and rapid response to prevent new populations from establishing [20]. In this chapter, I review the first of these three tenets, prevention, with a focus on terrestrial biosecurity programs for islands. Although I focus on terrestrial biosecurity, I acknowledge that the surrounding marine environment is just as important.

However, the complexity and distinct mechanisms driving marine invasions warrant separate, detailed exploration and are not covered here.

This review synthesizes information from peer-reviewed publications identified through targeted searches in the Web of Science and Scopus, in addition to gray literature identified through Google searches. Articles were prioritized for inclusion based on their relevance to preventative biosecurity measures implemented on island ecosystems. The findings were qualitatively assessed to highlight emerging themes and gaps in knowledge relating to island biosecurity preventative measures.

3. Prevention: Pre-border biosecurity

The invasive species curve demonstrates that prevention is the most cost-effective approach to managing invasive alien species [21]. Although preventative measures incur ongoing costs, these are substantially lower than the expenses involved in eradication programs. For example, in Mexico it has been shown to cost, on average, 20 times less to prevent the introduction of an invasive alien species than to carry out an eradication campaign [22]. Similarly, on South Georgia Island, the cost to build quarantine infrastructure to reduce the risk of rodent re-incursion was estimated to be 15 times less than the costs associated with a successful rodent eradication campaign [23]. In addition to being cost-effective, prevention further minimizes the impact of alien species on the island's ecosystem [3]. Whereas eradication campaigns can have significant non-target impacts, particularly where the application of toxins or non-target kill traps are used [24], preventative measures focus on reducing the risk of invasive species establishment and spread through early detection, thereby minimizing harm to non-target species and ecosystems.

Identifying pathways is a critical step in preventing the introduction of invasive alien species onto islands. Pathways refer to the processes or mechanisms by which species are transported from their native range to a new location [9]. Most introductions of alien species are the direct or indirect result of human activities. Globally, human activities have introduced more than 37,000 established alien species, of which 3500 are considered invasive [3]. Although human activity is often the primary cause of invasive alien species introductions, their subsequent spread is not always reliant on humans and needs to be taken into consideration in any biosecurity management plan. Natural dispersal mechanisms, such as wind, ocean currents, and animal movements, can also enable some species to reach new islands.

3.1 Human mediated intentional introductions

Many island communities depend heavily on the deliberate introduction of species [25]. These introductions can occur through agricultural, farming, the pet trade, or ornamental plant trade. Unfortunately, some species that are deliberately introduced may spread beyond their intended ranges and become invasive, subsequently having unwanted and unplanned effects on the island's economy or environment [26, 27]. On the Galapagos islands almost half of the 1500+ introduced alien species, were introduced intentionally, with many plant species still being intercepted despite now being prohibited [28]. For example, in 2016 the banana passionfruit (*Passiflora tripartite*) was intercepted 18 times. This species is not present on the Galapagos but is a known invasive alien species in Hawaii and New Zealand [28, 29]. Another plant that is widely traded through the tropics is the ivory cane palm (*Pinanga coronata*) [30]. This species was introduced to Fiji in the 1970s for its ornamental properties and has been identified as a potentially invasive alien plant species on Pacific and other oceanic islands [31]. Since its introduction into Fiji, the palm has impacted mahogany plantations [32] and native lowland rainforests [30]. Many ornamental and horticultural plants are referred to as "ticking time bombs," as they can take several decades [30, 31], or even up to 50 years, to establish and spread in the wild [33, 34].

The intentional introduction of plants can facilitate the unintentional introduction of other alien species including, insects, other plants, terrestrial invertebrates, vertebrates, and pathogens (e.g. [28]). On the Galapagos islands, most unintentional introductions arrived on introduced plants and plant associated material [28]. On Pityusic islands the ladder snake (*Rhinechis scalaris*) was introduced through the trade of ornamental olive trees originating from the Iberian Peninsula [35]. In the Phillipine Archipalego, the Asiatic painted toad (*Kaloula pulchra*) and greenhouse frog (*Eleutherodactylus planirostris*) have been transported through ornamental plants, nursery plants, and greenhouse material trades [36]. Many islands implement strict quarantine protocols and inspections for imported agricultural products to reduce the introduction of pests like fruit flies and plant pathogens.

The pet and farming trade are also major causes of introductions on islands. One of the primary pathways for introductions of reptiles is the pet trade, which is linked to many invasive species globally [37, 38]. For instance, the California kingsnake (*Lampropeltis californiae*) in the Canary Islands originated from captive populations [39, 40]. Similarly, the American bullfrog (*Lithobates catesbeianus*), Asiatic painted toad (*Kaloula pulchra*), and the Chinese bullfrog (*Hoplobatrachus rugulosus*), have all been introduced to the Phillipine Archipelago by the pet or farming trade [36], and goats have been introduced to many islands for their meat and milk production (e.g. [41]). On Christmas Island, many native species are listed as at risk from feral cat predation [42]. This has resulted in strict regulations for cat pet owners on the island, including registration and a ban on any further cat imports [42].

Although intentional human-mediated dispersal represents a substantial source of propagule pressure to islands, prevention measures are more manageable than for unintentional introductions. Management can be implemented through strict prohibition of introductions and the development of robust standards and regulations for animal and plant possession and trade. There are several established risk assessment methods available for screening species for deliberate introduction [43–45]. These risk assessments generally aim to evaluate risks associated with a specific alien species, commodity, or pathway (e.g. plants) to determine whether they should be accepted or rejected as an importation [46, 47]. These risk assessments can result in regulating the trade and breeding of these high-risk species, although there are often social challenges that need to be overcome when implementing these regulations, as illustrated by the case of banana passionfruit interceptions on the Galapagos islands, as outlined above. Striking a balance between protecting ecosystems and sustaining livelihoods is crucial for the long-term success of these biosecurity measures.

3.2 Human mediated unintentional introductions

The expansion of international trade has broadened the scope for human travel and maritime activity has led to greater transportation opportunities of hitchhiking species around the globe [48]. As islands are often dependent on shipping and air transport for supplies and trade, they are at significant biosecurity risk from unintentional human mediated introductions.

3.2.1 Vessels

Vessels, whether cargo ships, cruise ships, or private yachts, all have the potential to unintentionally transport terrestrial organisms (often referred to as topside). On Barrow island, an island off the coast of Western Australia, 28% of biosecurity risk detections were recorded on domestic vessel topsides, and 19% on international shipments [49], while on the Galapagos islands, 37 species (approximately 2% of all alien species detected on the Galapagos) were introduced via marine vessels [28]. Termites have been widely intercepted on vessels [50], while moths can be attracted to vessel lights while in port and inadvertently transported to new locations [51]. The Eurasian ladybird (*Coccinella undecimpunctata*) was likely introduced to South Georgia island as a stowaway on a ship [52]. The longest running, most extensive monitoring program of terrestrial arthropods undertaken on board a vessel detected 222 species (73,601 individual terrestrial arthropods), including the detection of four alien species, and one invasive alien species, *Tribolium destructor* [53]. Although, this chapter does not investigate marine introductions, it's worth noting that marine vessels are considered the primary pathway for introduced marine species through ballast water and hull fouling [54].

Some islands implement biosecurity procedures specifically designed for marine vessels, including topsides, to prevent pests from hitchhiking. For instance, Barrow Island has stringent biosecurity protocols for vessels accessing the Barrow Island Marine Terminal [55]. These procedures include detailed compliance inspections and strict housekeeping measures to ensure all topside areas are free of biosecurity risk materials, such as soil, plant matter, seeds, and fauna. Freshwater accumulation in receptacles on vessels is monitored to mitigate the risk of alien mosquito species, with any identified pooled water either drained, where possible, or treated. Waste management practices are enforced to prevent access by invertebrates and vertebrates, and while vessels are moored at Barrow Island, rodent guards are required to be installed on mooring lines, regularly monitored, and replaced if compromised. These guards, along with bait stations, often serve as critical mitigation strategies to minimize the risk of rodents reaching islands from marine vessels [23].

3.2.2 Air transport

Transit by air is generally more rapid than marine vessels and may therefore remove any barrier associated with hitchhiker survival due to shorter transit times [56]. On the Galapagos Islands, planes are a major pathway for insects and other terrestrial invertebrates, with large numbers of dead specimens being collected from the holds of planes post-fumigation during a trial conducted in 2008 [28]. A study completed at Miami International Airport in Florida (USA) found that approximately 10% of aircraft contained live insects in the cargo and interior following a random sampling of 1% of craft over a 1 year period (1998–1999) [57]. In Hawai'i, the Africanized Honey Bee (*Apis mellifera scutellate*), invasive ants (primarily the little fire ant (*Wasmannia auropunctata*) and the red imported fire ant (*Solenopsis invicta*)), coconut rhinoceros beetle (*Oryctes rhinoceros*) and mosquitos are targeted for surveillance at airports, primarily because of their ability to hitchhike undetected on planes [58]. In fact, mosquitoes are considered common hitchhikers on aircraft with a review by Mier-y-Teran-Romero, Tatem [59] of 17 studies (almost 600,000 aircraft) finding that approximately 0.91 mosquitoes were found on average per aircraft over a 68 year period. Despite many records of species hitchhiking on aircraft, direct reports of individual species being introduced and establishing on islands as a consequence of commercial air travel are uncommon. Although, the mosquito *Culicoides belkeni*, a pest species indigenous to Tahiti, reportedly successfully invaded other Polynesian islands after transport on aircraft [60].

3.2.3 Cargo

While sea vessels and aircraft can serve as direct vectors for invasive species, the people, their luggage, and other cargo carried by these vessels, represent a major biosecurity risk, and are known to transport a diverse assemblage of alien species [61]. Mice are habitual stowaways in cargo, with numerous records of their discovery in cargo destined for islands [62]. In New Zealand, there were 24 incidents of mice reaching islands among cargo within a seven-year period (2010–2017). The Asiatic painted toad (*Kaloula pulchra*) and cane toad (*Rhinella marina*) are considered high risk of incursion through cargo to the Phillipine Archipelago [36]. On St Helena, despite quarantine efforts, the ice plant (*Galenia papulose*) was found in river sand imported to the island for construction [8]. On Barrow Island, containerized goods account for 12% of biosecurity risk detections [49], however, due to the stringent biosecurity program implemented on Barrow Island this biosecurity risk material was detected and mitigated before it could establish [49].

To minimize biosecurity risks associated with cargo, robust biosecurity barriers should be established at critical points in the supply chain, such as during cargo preparation and fabrication [63–65]. For example, New Zealand has introduced pest-free warrants for businesses that service or supply goods to any island in Auckland's Hauraki Gulf [66, 67]. These warrants signify adherence to strict biosecurity standards, and encourage businesses to commit to providing pest-free goods and services, by promoting businesses with pest-free warrants as preferred suppliers for the islands [65]. Identifying biosecurity risk materials early in the supply chain ensures most risks are addressed before the cargo reaches the island, rather than during cargo inspections at the border. This alleviates operational pressures often associated with border cargo clearance. Furthermore, many islands do not have the facilities to mitigate border detections, resulting in cargo being sent off the island for treatment.

Innovative solutions are now emerging that will reduce biosecurity risk associated with this pathway. Dogs are now routinely used to check cargo for rodents before cargo departs Australia, during passage to, and on arrival at Macquarie Island [23, 68]. Design modifications are being trialed for shipping containers to reduce gaps and crevices available for hitchhikers [49, 69]. The Australian government has recently trialed an automated camera system combined with real-time machine learning to automatically scan the external surfaces of sea containers for contaminants and biosecurity risk material as they are being unloaded from cargo ships [70]. These innovations highlight the critical role of technology in strengthening biosecurity measures and mitigating the risks posed by global trade pathways.

3.2.4 Vehicles, machinery and equipment

The transport of vehicles, machinery, and equipment to islands, as well as their use on islands, can contribute to the introduction of alien species, in addition to facilitating their spread on islands. The main pest risk associated with vehicles are contamination with soil, invertebrates, plant debris, and seeds and other plant parts capable of propagation [71]. On St Helena island spiders were detected on 7% of imported vehicles (mostly cars), while 40% of vehicles were found to be contaminated with soil [8]. Nine percent of these soil samples were found to contain a mixture of grass and small dicotyledons that germinated following trials [8]. On the Canary Islands, a clear positive relationship between road kernel density estimates and invasive alien plant species was found, indicating the great influence that roads have on dispersal and establishment of invasive alien plants [72]. The invasion of the earthworm on San Clemente island, California (USA) is closely associated with roads, suggesting that earthworms are being dispersed by vehicles [73]. Dispersal of alien species, such as earthworms and plants, may be preventable by cleaning soil from tyre treads before moving vehicles between locations. The main phytosanitary measures applied to vehicles, machinery, and equipment are outlined in the International Standard for Phytosanitary Measures (ISPM) 41 [71].

3.2.5 Humanitarian and military operations

Humanitarian and military operations are considered higher-risk for biosecurity breaches due to their unique operational contexts, including rapid global deployments, visits to remote regions, and reduced biosecurity screening. For example, the brown tree snake (*Boiga irregularis*), introduced to Guam during World War II via U.S. military operations, has since been detected on other islands due to military aircraft movements from Guam; the snake has been intercepted on a military cargo plane in the Marshall Islands, Ryukya Islands, and has been repeatedly detected on Hawaii at the Hickam Air Force Base on the runway or aircraft hangers [74]. The incursion and establishment of two vertebrate invasive alien species, the common green iguana (*Iguana iguana*) and Cuba treefrog (*Osteopilus septentrionalis*), on Dominica Islands occurred as a result of post-hurricane relief supplies provided by multiple countries during the aftermath of Hurricane Maria [75]. Reduced biosecurity procedures during an emergency can inadvertently facilitate the introduction and establishment of invasive alien species by allowing species to bypass typical inspection and control measures. To prevent unintended movement of invasive alien species during critical recovery efforts, integrating biosecurity measures into disaster response frameworks should be considered.

3.2.6 People and luggage

People and their belongings are one of the most significant biosecurity risks associated with the accidental introduction of alien species to islands. On Barrow Island, nearly half of the 141 post-border alien species detections over a 6-year period were associated with the movement of people to the island [49]. McNeill, Phillips [76] surveyed soil removed from footwear carried in the baggage of international aircraft passengers arriving in New Zealand and found high incidences and diversities of viable bacteria, fungi, nematodes, seeds, and even live arthropods. Weed seeds are particularly adept at hitchhiking on footwear and clothing [77]. In Antarctica the flightless midge (*Eretmoptera murphyi*) is thought to be continuing its invasion of Signy island via treads of personnel boots [78]. In New Zealand, mice have been found in visitor daypacks, kayaks, and even nesting in dinghy flotation devices [62]. These examples highlight the critical need for preventative measures among travelers to address the risks posed by people and their belongings as vectors for invasive species.

Some islands have adopted extreme measures to safeguard their biodiversity from human impact, by regulating or restricting island access. On the Galapagos islands various systems have historically been implemented to regulate tourism. All visitors to protected areas must be accompanied by a tour guide, and group admission is regulated by a visitor management system to manage visitor numbers [79]. In 2022 further restrictions were recommended by the UNESCO World Heritage Committee to limit tourism expansion even further by including a construction monatorium and flight limits [79]. Although limiting the number of tourists on the Galapagos islands will certainly provide biosecurity benefits by reducing the risk of a biosecurity breach, these restrictions need to be carefully balanced against economic benefits of tourism to the island. On Barrow Island, public access is restricted, with only personnel working and residing on the island under a 'fly in, fly out' arrangement with access from two designated points of origin [80]. Restricted access has provided significant biosecurity benefits for Barrow Island by enabling strict regulation of inductions and ongoing awareness campaigns for all personnel traveling to the island. This approach has contributed to the workforce detecting over half of the post-border incursions on the island [49]. While controlled access may be an effective tool for mitigating biosecurity risks on some islands, it is not always appropriate, particularly for islands that are economically reliant on tourism, or inhabited islands. Islands are home to more than 600 million people globally [81], and the underappreciation of the social dimensions associated with island biosecurity remains a key challenge.

The concept of a 'shared responsibility' for biosecurity has gained traction over the last decade and is now increasingly incorporated into biosecurity management plans [82]. This concept emphasizes the collective action of all society in supporting biosecurity programs. However, fully embracing shared responsibility is hindered by a lack of investment in social psychology tools and an overreliance on biologists to influence resident and visitor behavior on islands [82]. To achieve the necessary reach and balance, biosecurity must effectively address human behavior as an integral part of its strategy [65]. There are many ways island biosecurity managers have tackled the challenges associated with a shared biosecurity program. The government of St Helena Island pursued a process of intensive consultation with stakeholders, which extended preparation of their biosecurity plan by about a year [8]. However, this active participation of all stakeholders was considered an important step in building a strong biosecurity culture among islanders and the social acceptance of the plan [8]. In contrast, other island managers develop a biosecurity plan, with a component of that plan involving outreach and culture building through various processes. For example, the implementation of Mexico's national strategy on invasive species is supported by fostering connections with island communities through outreach materials, educational talks, and creative workshops aimed at enhancing local identity by highlighting the island's biodiversity [22].

3.2.7 Risk assessments for unintentional introductions

Due to the finite resources often available to island biosecurity managers, it is not always possible to implement mitigation strategies across all potential pathways for alien species. Island managers are therefore often required to identify and prioritize the most significant species and pathways for intervention. Since gaining traction in biosecurity in the early 2000s, horizon scanning is now used to proactively identify biosecurity threats, prioritize species that are considered the highest risk for incursion and impact, and inform island mitigation strategies [61, 83–86]. For example, a horizon scan completed for the United Kingdom Overseas Territories, which consists of mostly islands that are widely distributed globally [61], identified most of the high risk species to be terrestrial invertebrates, with shipping containers, followed by ornamental plants, identified as the pathways associated with these species [61]. These results provide a foundation for the development of detailed pathway action plans for each island within the United Kingdom Overseas Territories.

There are a variety of predictive models employed by biosecurity managers that build on the insights from horizon scanning by quantitatively assessing the likelihood, pathways, and impacts of specific biosecurity incursions. These models provide detailed predictions that help prioritize resources and design targeted mitigation strategies. Examples include species distribution models (pest risk models) which can be used to predict where invasive alien species might arrive, establish, spread, or cause harmful impacts (e.g. [87]); climate modeling, which predicts how climatic factors (e.g., temperature, precipitation, wind patterns, and sea levels) influence the risk of introduction and establishment (e.g. [88–90]); and Bayesian Belief Networks, which assess pathway risks and probabilities [91]. This latter model has been applied to estimate the risk of 11 vertebrate invasive terrestrial species arriving, via five dispersal pathways, and establishing on 600 islands along the Pilbara coastline, Western Australia. The number of visitors to these islands was found to greatly influence the number of animals arriving with one arrival per 300 visitors predicted across the 11 species studied [91]. Unfortunately, a lot of the data required for these types of predictive models are lacking in the scientific literature, resulting in expert elicitation being used to fill information gaps. However, by integrating a combination of horizon scanning with predictive models, biosecurity managers can develop a more comprehensive and proactive management plan.

3.3 Natural dispersal

For millennia, natural dispersal has shaped the biogeography and unique biodiversity of oceanic islands, which have never been connected to continental landmasses [92]. However, when alien species are introduced via human pathways and then spread further using natural dispersal, the rate of expansion can far outpace natural processes. This dual pathway creates significant biosecurity challenges, as human activity not only circumvents natural barriers, but also facilitates the establishment of invasive species, which can then exploit natural dispersal to widen their range. Effective biosecurity measures must therefore address both the prevention of initial introductions and the management of secondary spread.

Some invasive alien species are known to colonize new islands by swimming. Dispersal by swimming is primarily restricted to invasive vertebrates, with species exhibiting varying swimming capabilities. Although scientific data on the natural dispersal capability of alien invasive mammals is limited, there are several records of invasive vertebrates establishing on islands by swimming. Stoats have made their way to many islands in New Zealand via swimming [13, 67, 93], while red deer established on Secretary Island after swimming from the South Island of New Zealand in the early 1960s [94]. Mice reinvaded Pomona and Rona Islands in Lake Manapouri within a decade of successful eradication [62]. These islands are 500 and 600 m offshore respectively, with the re-invasions coinciding with beech masting events when mice reached high abundance on the mainland [62]. For islands that are not well isolated, unassisted introduction by invasive alien vertebrates can represent a significant risk.

Some invasive alien species can disperse long distances from their entry points via flight. For example, the great white butterfly, *Pieris brassicae*, can migrate hundreds of kilometers to new locations within a season [95], while the European starling (*Sturnus vulgaris*) can disperse up to 1000 km [96]. Wind promotes incursions of the island sugarcane plant-hopper *Eumetopina flavipes* from Papua New Guinea to the Torres Strait islands [97], while seeds of the invasive alien plant ivory cane palm (*Pinanga coronata*) are thought to be dispersed by birds on Fiji [32]. Alien species have also been recorded using islands of Torres Strait as 'stepping stones' to cross the 150 km of shallow sea that separates Papua New Guinea from mainland Australia [98].

Extreme weather conditions can further facilitate the dispersal of alien species at regional and global scales. For example, the cactus moth (*Cactoblastis cactorum*) was likely blown from host islands in the Caribbean to Mexico during the 2005 hurricane season [99]. Rats established on Green Cay in the Caribbean following the 2017 hurricane, most likely through rafting on storm debris, swimming, or a combination of both [100]. The red palm mite (*Raoiella indica*) has been spread throughout the Caribbean islands by major storms and hurricanes [99], while the non-native balloon vine (*Cardiospermum grandiflorum*) and mile-a-minute vine (*Mikania micrantha*) have proliferated on part of the Cook islands by seeds blown by Cyclone Sally [99]. The frequency of these extreme weather events are likely to increase with climate change [99].

4. Island biosecurity governance

History has shown that island biosecurity programs work better when there is a single governing agency. For example, the Channel Islands encompass eight islands ranging in size from 250 to 25,000 ha, as well as numerous surrounding islets. Even though discrete islands within the archipelago are managed by various jurisdictions, the landowning entities jointly fund a full-time position to lead the implementation of group biosecurity objectives [101]. This coordinated and collaborative approach to biosecurity has provided numerous efficiencies and enhanced the effectiveness of the program [11, 101]. On the Galapagos islands, centralisation of regional biosecurity management in 2012 under one organization, the Agency for the Regulation and Control of Biosecurity and Quarantine for Galapagos (ABG), has been significant in improving the efficiency and scope of surveillance for current and emerging threats.

Although collaboration at the local and national level has been demonstrated to be important, invasive species are an international problem. For the effective biosecurity management of islands, coordinated action by international government agencies is required [9]. In North America, island biosecurity is a key focus of the Trilateral Committee for Wildlife and Ecosystem Conservation and Management [102]. The purpose of this committee, and subsequent agreements, are to align three countries (Canada, Mexico, and the United States) in a continental endeavor to safeguard North American wildlife and ecosystems. The resulting collaborative efforts has resulted in many success stories, including the restoration of seabird habitat in Gwaii Haanas by the eradication of invasive black rats, and the eradication of feral sheep from Socorro Island [103]. In the Pacific region, the *Guidelines for Invasive Species Management cover the Pacific Island Countries and Territories* (PICTs), which include 22 nations and territories in the region [19]. The guidelines focus on regional collaboration and capacity-building for addressing island invasive species threats to biodiversity, ecosystems, and livelihoods across these islands. These guidelines have fostered regional cooperation through mechanisms like the Pacific Regional Invasive Species Management Support Service (PRISMSS), and the inaugural Pacific Ecological Security Conference (PESC) held in 2022 [104]. The PESC convened over 100 island leaders, policy-makers, natural resource managers and global and regional invasive species experts to address critical biosecurity issues for the Pacific. These mechanisms are essential for enabling countries to share expertise, resources, and strategies.

The success of the *Pacific Guidelines for Invasive Species Management* has led to their adaptation and update for global use in island biosecurity management. The updated version, now known as the *Guidelines for Invasive Species Planning and Management on Islands* [9], not only incorporate lessons learnt from the Pacific region, but also align with international frameworks such as the Convention on Biological Diversity and the Aichi targets for biodiversity conservation. Providing practical recommendations for developing and implementing effective biosecurity strategies on islands to prevent, control, or eradicate invasive species these guidelines have been widely adopted by island managers across the globe and have played a crucial role in increasing collaboration among stakeholders, including governments, conservation groups, and local communities. Although not island specific, there are now a diversity of international obligations and agencies linked to biosecurity, including World Trade Organization agreements, the International Plant Protection Convention, the World Organization for Animal Health, and the World Health Organization [82]. Despite increased awareness of the importance and seriousness of invasive species, there continues to be an urgent need for increased and coordinated action at the local, national and regional levels to address biosecurity globally.

5. Conclusions

> *'Biosecurity is a casualty of its own success. When everything works as it should, nothing happens.' [65]*

This phrase highlights a paradox inherent in effective biosecurity systems. Biosecurity preventative measures often go unnoticed compared to the more expensive reactive responses required once an alien invasive species is detected on an island. When preventative measures function well, there are no visible crises to draw attention to the importance of biosecurity. This can lead to underappreciation, complacency, or reduced funding and support, as it may appear that risks were never significant in the first place. Essentially, the absence of alien species incursions, due to effective preventative measures, can make its value less obvious to stakeholders and decision-makers. The challenge therefore lies in ensuring ongoing support for preventative biosecurity efforts despite the lack of immediate, tangible results.

Author details

Melissa L. Thomas
Harry Butler Institute, Murdoch University, Perth, Australia

*Address all correspondence to: melissa.thomas@murdoch.edu.au

References

[1] UN-OHRLLS. Small Island Developing States in Numbers: Biodiversity and Oceans. Office of the High Representative for the Least Developed Countries, Landlocked Developing Countries and Small Island Developing States (UN-OHRLLS); 2017. Available from: https://www.un.org/ohrlls/

[2] Kier G et al. A global assessment of endemism and species richness across island and mainland regions. Proceedings of the National Academy of Sciences of the United States of America. 2009;**106**(23):9322-9327

[3] Roy HE et al. In: Genovesi P, Wilson JR, editors. Summary for Policymakers of the Thematic Assessment of Invasive Alien Species and their Control of the Intergovernmental Platform on Biodiversity and Ecosystem Services - Advanced Unedited Version. Bonn, Germany: Intergovernmental Science-Policy Platform on Biodiversity and Ecosystem Services; 2023. p. 41

[4] Loehle C, Eschenbach W. Historical bird and terrestrial mammal extinction rates and causes. Diversity and Distributions. 2012;**18**(1):84-91

[5] Keitt B et al. The global islands invasive vertebrate eradication database: A tool to improve and facilitate restoration of island ecosystems. In: Veitch CR, Clout MN, Towns DR, editors. Island Invasives: Eradication and Management. Gland, Switzerland: IUCN; 2011

[6] Bellard C, Cassey P, Blackburn TM. Alien species as a driver of recent extinctions. Biology Letters. 2016;**12**(2)

[7] Doherty TS et al. Invasive predators and global biodiversity loss. Proceedings of the National Academy of Sciences of the United States of America. 2016;**113**(40):11261-11265

[8] Balchin JR et al. Biosecurity on St Helena island – A socially inclusive model for protecting small island nations from invasive species. In: Veitch CR et al., editors. Island Invasives: Scaling up to Meet the Challenge. Gland, Switzerland: IUCN; 2019. pp. 468-472

[9] IUCN. In: Tye A, editor. Guidelines for Invasive Species Planning and Management on Islands. UK and Gland, Switzerland; 2018. p. 52

[10] Chao RF, Lin TE. Effect of citizen action on suppression of invasive alien lizard populations: A case of the removal of Eurtropis multifasciata on Green Island, Taiwan. Applied Ecology and Environmental Research. 2017;**15**(2):1-13

[11] Matos J et al. Connecting island communities on a global scale: Case studies in island biosecurity. Western North American Naturalist. 2019;**78**(4):959-972

[12] Liu Y et al. Safeguarding Island Ecosystems: A Review of Island Global Biosecurity Plans. Western Australia. 2025

[13] Auld M et al. Safeguarding Orkney's native wildlife from non-native invasive stoats. In: Veitch CR et al., editors. Island Invasives: Scaling up to Meet the Challenge. Gland, Switzerland: IUCN; 2019. pp. 244-248

[14] Siers SR et al. Limitations of invasive snake control tools in the context of a new invasion on an island with abundant prey. Neobiota. 2024;**90**:1-33

[15] Siers SR et al. In situ evaluation of an automated aerial bait delivery system for landscape-scale control of invasive brown treesnakes on Guam. In: Veitch CR et al., editors. Island Invasives: Scaling up to Meet the Challenge. Gland, Switzerland: IUCN; 2019. pp. 348-355

[16] DBCA. Houtman Abrolhos Islands Reserve Management Plan 2022-2032. Kensington WA: Department of Primary Industries and Regional Development; 2022. p. 110

[17] Thomas ML, Gray M, Campos M. Houtman Abrolhos Islands National Park: Tererstrial Biosecurity Guidelines. Harry Butler Institute, Murdoch University. Report prepared for the Department of Biodiversity, Conservation and Attractions, Western, Australia: Western Australia; 2022. p. 108

[18] DOC. Island Biosecurity Best Practice Guidelines. New Zealand: Department of Conservation; 2020. p. 80

[19] SPREP. Guidelines for Invasive Species Management in the Pacific: A Pacific Strategy for Managing Pests, Weeds and Other Invasive Species. Apia, Samoa: S.o.t.P.R.E. Programme; 2009

[20] Russell JC et al. Testing island biosecurity systems for invasive rats. Wildlife Research. 2008;**35**(3):215-221

[21] Hobbs RJ, Humphries SE. An integrated approach to the ecology and management of plant invasions. Conservation Biology. 1995;**9**(4):761-770

[22] Latofski-Robles M et al. Mexico's island biosecurity programme: Collaborative formulation and implementation. In: Veitch CR et al., editors. Island Invasives: Scaling up to Meet the Challenge. Gland, Switzerland: IUCN; 2019. pp. 484-488

[23] Richardson MG, Croxall JP. Achieving post-eradication biosecurity on South Georgia. In: Veitch CR et al., editors. Island Invasives: Scaling up to Meet the Challenge. Gland, Switzerland: IUCN; 2019. pp. 489-493

[24] Carey PW. Simultaneous rat, mouse and rabbit eradication on Bense and Little Bense Islands, Falkland Islands. In: Veitch CR et al., editors. Island Invasives: Scaling up to Meet the Challenge. Gland, Switzerland: IUCN; 2019. pp. 108-113

[25] Ewel JJ et al. Deliberate introductions of species: Research needs – Benefits can be reaped, but risks are high. Bioscience. 1999;**49**(8):619-630

[26] Richardson DM, Higgins SI. Pines as invaders in the southern hemisphere. In: Richardson DM, editor. Ecology and Biogeography of Pinus. Cambridge: Cambridge University Press; 1998. pp. 450-473

[27] Virtue JG, Bennett SJ, Randall RP. Plant introductions in Australia: How can we resolve 'weedy' conflicts of interest. In: Proceedings of the 14th Australian Weeds Conference. Wagga, NSW: Weed Society of New South Wales; 2004

[28] Toral-Granda MV et al. Alien species pathways to the Galapagos Islands, Ecuador. PLoS One. 2017;**12**(9):e0184379. DOI: 10.1371/journal.pone.0184379

[29] Beavon MA, Kelly D. Invasional meltdown: Pollination of the invasive liana Passiflora tripartita var. mollissima (Passifloraceae) in New Zealand. New Zealand Journal of Ecology. 2012;**36**(1):100-107

[30] Keppel G, Watling D. Ticking time bombs – Current and potential future impacts of four invasive plant species on the biodiversity of lowland tropical rainforests in South-East Viti Levu, Fiji.

The South Pacific Journal of Natural and Applied Sciences. 2011;**29**:42-45

[31] Meyer J-Y, Lavergne C, Hodel DR. Time bombs in gardens: Invasive ornamental palms in tropical islands, with emphasis on French Polynesia (Pacific Ocean) and the Mascarenes (Indian Ocean). Palms. 2008;**52**:71-83

[32] Dyer MJB et al. Using expert knowledge and field surveys to guide management of an invasive alien palm in a Pacific Island lowland rainforest. In: Veitch CR et al., editors. Island Invasives: Scaling up to Meet the Challenge. Gland, Switzerland: IUCN; 2019. pp. 417-423

[33] Sullivan JJ, Timmins SM, Williams PA. Movement of exotic plants into coastal native forests from gardens in northern New Zealand. New Zealand Journal of Ecology. 2005;**29**(1):1-10

[34] Sullivan JJ et al. People and time explain the distribution of naturalized plants in New Zealand. Weed Technology. 2004;**18**:1330-1333

[35] Picó G et al. Control of the ladder snake (Rhinechis scalaris) on Formentera using experimental live-traps. In: Veitch CR et al., editors. Island Invasives: Scaling up to Meet the Challenge. Gland, Switzerland: IUCN; 2019. pp. 332-336

[36] Pili AN et al. Spatial dynamics of invasion and distribution of alien frogs in a biodiversity hotspot archipelago. In: Veitch CR et al., editors. Island Invasives: Scaling up to Meet the Challenge. Gland, Switzerland: IUCN; 2019. pp. 337-347

[37] Krysko KL et al. New verified nonindigenous amphibians and reptiles in Florida through 2015, with a summary of over 152 years of introductions. International Reptile Conservation Foundation Reptiles and Amphibians. 2016;**23**:110-143

[38] McFadden MS, Topham P, Harlow PS. A ticking time bomb: Is the illegal pet trade a pathway for the establishment of corn Snake (Elaphe guttata) populations in Australia? Australian Zoologist. 2017;**38**:499-504

[39] Fisher SR et al. Life-history comparisons between the native range and an invasive island population of a colubrid snake. In: Veitch CR et al., editors. Island Invasives: Scaling up to Meet the Challenge. Gland, Switzerland: IUCN; 2019. pp. 326-331

[40] Fisher S et al. Reproductive plasticity as an advantage of snakes during island invasion. Conservation Science and Practice. 2021;**3**(12):e554. DOI: 10.1111/csp2.554

[41] Algar D et al. Dirk Hartog Island 'return to 1616' project – The first six years (2014 to 2019). Ecological Management and Restoration. 2020;**21**(3):173-183

[42] Algar D, Hamilton N, Pink C. Progress in eradicating cats (Felis catus) on Christmas Island to conserve biodiversity. Raffles Bulletin of Zoology. 2014;**30**:45-53

[43] Lehtonen PP. Pest risk assessment in the United States: Guidelines for qualitative assessments for weeds. In: Groves RHPF, Virtue JG, editors. Weed Risk Assessment. Melbourne: CSIRO Publishing; 1996. pp. 117-123

[44] Walton CS. Implentation of a permitted list approach to plant introductions to Australia. In: RH PFG, Virtue JG, editors. Weed Risk Assessment for Plant Introductions to Australia. Melbourne: CSIRO Publishing; 1996. pp. 93-99

[45] Williams PA, Nicol E, Newfiled M. Assessing the risk to indigenous biota

of plant taxa new to New Zealand. In: RH PFG, Virtue JG, editors. Weed Risk Assessment for Plant Introductions to Australia. Melbourne: CSIRO Publishing; 1996. pp. 100-116

[46] Hulme PE. Weed risk assessment: A way forward or a waste of time? Journal of Applied Ecology. 2012;**49**(1):10-19

[47] Pheloung PC, Williams PA, Halloy SR. A weed risk assessment model for use as a biosecurity tool evaluating plant introductions. Journal of Environmental Management. 1999;**57**(4):239-251

[48] Banks NC et al. The role of global trade and transport network topology in the human-mediated dispersal of alien species. Ecology Letters. 2015;**18**(2):188-199

[49] Scott JK, McKirdy SJ, van der Merwe J. et al. Zero-tolerance biosecurity protects high-conservation-value island nature reserve. Scientific Reports. 2017;**7**:772. DOI: 10.1038/s41598-017-00450-y

[50] Scheffrahn RH, Crowe W. Ship-borne termite (Isoptera) border interceptions in Australia and onboard infestations in Florida, 1986-2009. Florida Entomologist. 2011;**94**(1):57-63

[51] Barnes DKA, Convey P. Odyssey of stow-away noctuid moths to southern polar islands. Antarctic Science. 2005;**17**(3):307-311

[52] Tichit P et al. First record of the introduced ladybird beetle, Coccinella undecimpunctata Linnaeus (1758), on South Georgia (sub-Antarctic). Ecology and Evolution. 2023;**13**(9). DOI: 10.1002/ece3.10513

[53] McKirdy SJ, O'Connor S, Thomas ML, et al. Biosecurity risks posed by a large sea-going passenger vessel: Challenges of terrestrial arthropod species detection and eradication. Scientific Reports. 2019;**9**:19339. DOI: 10.1038/s41598-019-55554-4

[54] Bax N et al. Marine invasive alien species: A threat to global biodiversity. Marine Policy. 2003;**27**(4):313-323

[55] Chevron. Quarantine Procedure Trading Vessels. Vol. 11. Perth, Australia: Chevron; 2024

[56] Hughes KA et al. Impact of anthropogenic transportation to Antarctica on alien seed viability. Polar Biology. 2010;**33**(8):1125-1130

[57] Dobbs TT, Brodel CF. Cargo aircraft as a pathway for the entry of nonindigenous pests into South Florida. Florida Entomologist. 2004;**87**(1):65-78

[58] Kaufman LV. Mamalu Poepoe: Enhancing Hawaii's biosecurity through interagency collaboration. Annals of the Entomological Society of America. 2021;**114**(6):694-701

[59] Mier-y-Teran-Romero L, Tatem AJ, Johansson MA. Mosquitoes on a plane: Disinsection will not stop the spread of vector-borne pathogens, a simulation study. PLoS Neglected Tropical Diseases. 2017;**11**(7).e0005683. DOI: 10.1371/journal.pntd.0005683

[60] Lounibos LP. Invasions by insect vectors of human disease. Annual Review of Entomology. 2002;**47**:233-266

[61] Dawson W et al. Horizon scanning for potential invasive non-native species across the United Kingdom overseas territories. Conservation Letters. 2022;**16**(1)e12928. DOI: 10.1111/conl.12928

[62] Broome K et al. House mice on islands: Management and lessons from

New Zealand. In: Veitch CR et al., editors. Island Invasives: Scaling up to Meet the Challenge. Gland, Switzerland: IUCN; 2019. pp. 100-107

[63] Chevron. Barrow Island Quarantine: Beyond Best Practice. 2015. p. 6

[64] Morgan S et al. An Assessment of Risk Pathways from Commercial Businesses to Pest-Free Islands of the Hauraki Gulf. New Plymouth, New Zealand: New Zealand Biosecurity Institute; 2014

[65] Kennedy ES, Broome KG. How do we prevent the obstacles to good island biosecurity from limiting our eradication ambitions? In: Veitch CR et al., editors. Island Invasives: Scaling up to Meet the Challenge. Gland, Switzerland: IUCN; 2019. pp. 478-483

[66] Auckland Council. 1 January 2024 Cutoff Date to be Pest Free Warranted. 2023. Available from: https://ourauckland.aucklandcouncil.govt.nz/news/2023/11/pest-free-warrant/

[67] Bassett IE et al. Treasure islands: Biosecurity in the Hauraki Gulf marine park. New Zealand Journal of Ecology. 2016;**40**(2):250 266

[68] Moran J, McLennan A. Meet Nui, the little dog with the big job – Keeping world heritage-listed Macquarie Island pest free. Tasmania, Australia: Australian Broadcasting Coorporation News. 2020. ABC: Online

[69] Karri R. Sea container structures – Risks. In: International Symposium: Optimising Sea Container Designs. Rotterdam. International Plant Protection Convention; 2024. Available from: https://assets.ippc.int/static/media/files/publication/en/2024/11/Sea_Container_Structures_-_Risks.pdf

[70] DAFF, 96-2022: Upcoming Biosecurity Automated Threat Detection System (BATDS) Trial at DP World Australia's Brisbane Terminal. Western Australia: Department of Agriculture, Forestry and Fisheries; 2022. Available from: https://www.agriculture.gov.au/biosecurity-trade/import/industry-advice/2022/96-2022

[71] International Plant Protection Convention. International Standard for Phytosanitary Measures No. 41: International Movement of Used Vehicles, Machinery and Equipment. Rome, Italy: Food and Agriculture Organization of the United Nations; 2017. p. 14

[72] Da Re D et al. Modelling invasive plant alien species richness in Tenerife (Canary Islands) using Bayesian Generalised Linear Spatial Models. In: Veitch CR et al., editors. Island Invasives: Scaling up to Meet the Challenge. Gland, Switzerland: IUCN; 2019. pp. 410-416

[73] Paudel S et al. Predicting spatial extent of invasive earthworms on an oceanic island. Diversity and Distributions. 2016;**22**(10):1013-1023

[74] USGS. Rown Tree Snake Colonization and Dispersal Events on Guam. 2024. Available from: file:///Users/melissathomas/Downloads/report/invasion/colonize.html

[75] van den Burg MP, Brisbane JLK, Knapp CR. Post-hurricane relief facilitates invasion and establishment of two invasive alien vertebrate species in the Commonwealth of Dominica, West Indies. Biological Invasions. 2020;**22**(2):195-203

[76] McNeill M et al. Transportation of nonindigenous species via soil on international aircraft passengers'

footwear. Biological Invasions. 2011;**13**(12):2799-2815

[77] Ansong M, Pickering C. Weed seeds on clothing: A global review. Journal of Environmental Management. 2014;**144**:203-211

[78] Bartlett JC et al. The effectiveness of Virkon® S disinfectant against an invasive insect and implications for Antarctic biosecurity practices. Antarctic Science. 2021;**33**(1):1-9

[79] Vega-Mariño PA et al. Review of one health in the Galápagos Islands (part 2): Climate change, anthropogenic activities, and socioeconomic sustainability. Frontiers in Conservation Science. 2024;**5**. DOI: 10.3389/fcosc.2024.1351716

[80] Thomas ML et al. Many eyes on the ground: Citizen science is an effective early detection tool for biosecurity. Biological Invasions. 2017;**19**(9):2751-2765

[81] GLISPA. Island Biodiversity - Island Bright Spots in Conservation and Sustainability. S.o.t.C.o.B. Diversity, Editor; 2014

[82] Rawluk A, Beilin R, Lavau S. Enacting shared responsibility in biosecurity governance: Insights from adaptive governance. Ecology and Society. 2021;**26**(2). DOI: 10.5751/ES-12368-260218

[83] Hughes KA et al. Invasive non-native species likely to threaten biodiversity and ecosystems in the Antarctic Peninsula region. Global Change Biology. 2020;**26**(4):2702-2716

[84] Lieurance D et al. Identifying invasive species threats, pathways, and impacts to improve biosecurity. Ecosphere. 2023;**14**(12)

[85] Lucy FE et al. Horizon scan of invasive alien species for the island of Ireland. Management of Biological Invasions. 2020;**11**(2):155-177

[86] Peyton JM et al. Horizon scanning to predict and prioritize invasive alien species with the potential to threaten human health and economies on cyprus. Frontiers in Ecology and Evolution. 2020;**8**

[87] Venette RC et al. Pest risk maps for invasive alien species: A roadmap for improvement. Bioscience. 2010;**60**(5):349-362

[88] Harris RJ, Barker G. Relative risk of invasive ants (Hymenoptera: Formicidae) establishing in New Zealand. New Zealand Journal of Zoology. 2007;**34**(3):161-178

[89] Chen YH. Global potential distribution of an invasive species, the yellow crazy ant (*Anoplolepis gracilipes*) under climate change. Integrative Zoology. 2008;**3**(3):166-175

[90] Perry G, Vice D. Forecasting the risk of brown tree snake dispersal from Guam: A mixed transport-establishment model. Conservation Biology. 2009;**23**(4):992-1000

[91] Lohr C et al. Predicting island biosecurity risk from introduced fauna using Bayesian Belief Networks. Science of the Total Environment. 2017;**601**:1173-1181

[92] Cowie RH, Holland BS. Dispersal is fundamental to biogeography and the evolution of biodiversity on oceanic islands. Journal of Biogeography. 2006;**33**(2):193-198

[93] King CM et al. Swimming capabilities of stoats and the threat to inshore sanctuaries. Biological Invasions. 2014;**16**(5):987-995

[94] Macdonald N et al. Eradication of red deer from Secretary Island, New Zealand: Changing tactics to achieve success. In: Veitch CR et al., editors. Island Invasives: Scaling up to Meet the Challenge. Gland, Switzerland: IUCN; 2019. pp. 256-260

[95] Spieth HR, Cordes R. Geographic comparison of seasonal migration events of the large white butterfly, *Pieris brassicae*. Ecological Entomology. 2012;**37**(6):439-445

[96] Stuart KC et al. Global invasion history and native decline of the common starling: Insights through genetics. Biological Invasions. 2023;**25**(5):1291-1316

[97] Anderson KL et al. Wind-assisted migration potential of the island sugarcane planthopper Eumetopina flavipes (Hemiptera: Delphacidae): Implications for managing incursions across an Australian quarantine frontline. Journal of Applied Ecology. 2010;**47**(6):1310-1319

[98] Davis RI et al. Northern Australia quarantine strategy plant health surveys: Over thirty years of a globally unique on- and off-shore solution to island nation biosecurity challenges. Historical Records of Australian Science. 2024;**35**(2):223-234

[99] Burgiel SW, Muir AA. Invasive Species, Climate Change and Ecosystem-Based Adaptation: Addressing Multiple Drivers of Global Change. Washington, DC, US, and Nairobi, Kenya: Global Invasive Species Programme; 2010

[100] Shiels AB et al. Invasive rat establishment and changes in small mammal populations on Caribbean Islands following two hurricanes. Global Ecology and Conservation. 2020;**22**:e00986. DOI: 10.1016/j.gecco.2020.e00986.

[101] Boser CL et al. Proactive planning and compliance for a high-priority invasive species rapid response programme. In: Veitch CR et al., editors. Island Invasives: Scaling up to Meet the Challenge. Gland, Switzerland: IUCN; 2019. pp. 473-477

[102] TCWECM. Trilateral Committee for Wildlife and Ecosystem Conservation and Management. Facilitating international cooperation for conserving the living heritage of North America. Canada, Mexico, US; 2024. Available from: https://trilat.org/

[103] TIIWG. In: T.I.I.W. Group, editor. 2017 Summary Report. Sheperdstown, WV; 2018. p. 11

[104] Hoffmann BD et al. Three new strategies for improving biosecurity and invasive species management to build resilience in Pacific Islands. Neobiota. 2024;**92**:193-210

Chapter 3

How Marine Plastic Pollution Education Develops Secondary School Students' Sustainability Competences

Viktoria Smenes and Hilde Ervik

Abstract

Marine plastic pollution is a global environmental challenge. Providing marine plastic pollution education is essential for combating marine plastic pollution. Knowledge about biodiversity and an understanding of how biodiversity should be safeguarded should be included in education on marine plastic pollution. This study provides answers about the knowledge, understanding and attitude that Norwegian secondary school students (n = 50) have about how marine plastic pollution affects biodiversity and how biodiversity can be protected. The results from the questionnaires (pre-test and post-test) and group interviews showed that after having completed their education, over half (51.4%) of the respondents possessed adequate knowledge of and a good understanding of how biodiversity can be safeguarded and 29.7% showed a good understanding of why microplastics pose a threat to biodiversity. 35.1% provided good arguments as to why they believe their actions will make a difference. However, the results were less promising with respect to how the international community should cooperate on sustainable development. This study provides a foundation for what should be included in the education of secondary school students on marine plastic pollution, and how it should be provided.

Keywords: education for sustainable development (ESD), action competences, sustainability, biodiversity, macro and microplastics, field and laboratory work, marine littering

1. Introduction

Sustainable development refers to development that covers current requirements without compromising the requirements of future generations [1]. In Ref. [2], it is pointed out that the term sustainability is often incorrectly applied, especially when it is limited to environmental issues without taking into account the social and economic aspects. According to Ref. [2], a fragmented approach has contributed to a lack of agreement on how sustainability competencies should be defined and taught.

IntechOpen

In order to solve sustainability challenges, create a sustainable future and promote social, economic and environmental sustainability, students must acquire knowledge, attitudes, skills and values from Education for Sustainable Development (ESD) [3, 4], and there is currently a crucial need for action [2] if we are going to be able to meet our ever-increasing sustainability challenges. Schools, with their mandates, play a critical role in promoting ESD, because the current climate and environmental crisis requires the way in which we educate future generations to be restructured [5]. Education must go beyond traditional academic disciplines and focus more on the development of competencies that will enable students to face complex global challenges [6, 7]. This will lay the foundations for a more sustainable future [5].

Sustainability competencies, as discussed in several studies [2, 6, 8, 9], include skills such as systems thinking, future thinking, critical reflection, cooperation and the ability to act responsibly. These skills are essential for understanding and addressing the complex challenges associated with sustainability dilemmas. In Ref. [2], it is emphasised that ESD is a key component in developing such competencies, and the overall perspective on sustainable competences is regarded as being key for equipping future generations to become active change agents.

Several studies have pointed out that education on sustainable development has a positive impact on students' attitudes and behaviour with respect to sustainability because it raises awareness about environmental challenges and also strengthens students' critical thinking about complex sustainability issues [10]. A Swedish longitudinal study shows that ESD increases students' competence to act, provides them with the skills to tackle sustainability challenges and helps them to take action in their everyday lives [11]. This serves to emphasise the fact that ESD has a long-term impact on students' ability to contribute to sustainable solutions. Ref. [12] describes a study conducted at three upper secondary schools in Norway which emphasises that school experiences that allow students to positively influence society, correlate with their competence to act and that this competence includes knowledge, self-confidence and a willingness to act. The findings also indicated that providing students with the opportunity to influence society can promote their development of sustainability action competence [12]. Ref. [3] explains that key issues with respect to sustainable development, for example, biodiversity and climate change, must be included in teaching and learning. Students should participate actively in order to be empowered and motivated to implement sustainable development measures, and learning and teaching methods are required in order to facilitate this [3].

Society is dependent on having a sustainable coexistence with the planet [2]. Every year, 19–23 million tonnes of plastic end up in the ocean, threatening biodiversity and amplifying climate change [13, 14]. About 88% of marine species are negatively affected by plastic [15], and plastic waste is broken down into microplastics which damage organisms and the environment [16]. Plastic pollutants exacerbate the stressors created by climate change and affect species and public health [17–19]. According to Ref. [19], several solutions are required, because the causes are diverse and complex. The solution categories outlined in Ref. [19] are circular economy, awareness/bans/less consumption, price-fixing/fees/subsidies, better raw materials/recycling and reuse and clean-up/better waste management. Measures must be implemented at several levels of society, and Ref. [20] shows that there is no stronger motivator than education for promoting measures designed to combat biodiversity loss. It is believed that education can help students understand the consequences of what marine plastic waste entails and the lessons learnt at school will influence students' knowledge, understanding and attitudes about marine plastic pollution [20–22]. Knowledge must be developed along with skills, attitudes and values in order to effect behavioural changes [23], such as increased awareness, less consumption, recycling and cleaning up. Being involved and implementing measures will create the belief that making a difference works and will create hope for the future because involvement is positively related to hope [24, 25].

Marine plastic pollution is a complex sustainability dilemma, and the academic context included in teaching may vary [21]. Several studies have examined students' knowledge, skills, attitudes and values with respect to various sustainability dilemmas, including nature conservation [26–29]. However, more research is required about the knowledge, understanding and attitudes that students have about how marine plastic pollution affects biodiversity and how biodiversity can be safeguarded. In Ref. [21], it is emphasised that more research is required in order to ensure quality assurance practices with respect to teaching marine pollution. Teacher education should include more focus on marine littering in order to equip teachers with the knowledge and skills required for teaching this subject in a suitable way [21]. Ref. [21] states that teachers must have a good, thorough understanding of the problem and also be able to pass on such knowledge effectively to their students. It was also discovered that teachers' attitudes towards marine littering and how they teach such are important factors with respect to the benefits of teaching marine littering.

This study examined the sustainability competencies developed by Norwegian secondary school students when undertaking a modelled sustainability project, which was designed to enable students to understand how marine plastic pollution affects biodiversity. The findings provide a basis for how marine plastic pollution education can be carried out. We start below by explaining the conceptual background and literature underlying the study, followed by the methods, results, discussion and conclusions.

2. Conceptual background

The study was originally motivated by the first author's own desire to design a teaching programme for secondary school students, aimed at being able to contribute towards students' understanding of how to maintain biodiversity, based on the threat to nature imposed by marine plastic pollution. In preparation for designing this study, the first author participated in cleaning up marine waste, along with coastal renovators from Eider AS Mausund Feltstasjon[1], in order to gain an insight into the extent of marine plastic littering on the Froan archipelago in central Norway.

Schools play a key role; teachers are important change agents with respect to teaching about marine plastic pollution [21], and students will be made aware of how plastic waste needs to be handled by acquiring increased knowledge [30]. The teaching provided not only includes the incorporation of sustainability competencies but also embraces innovative and forward-thinking approaches that motivate and empower students to change their behaviour and provide them with the desire to take action that will ensure a better future [23, 30]. Studies conducted with groups of students similar to those in this case study have examined how teaching can increase

[1] www.eider.no

students' awareness and actions, focusing particularly on their knowledge, attitudes and skills with respect to plastic pollution. The results indicate the suitability of conducting practical work in lessons on plastic littering. Ref. [30] explains that practical teaching strategies improved students' understanding of plastic waste problems and handling, while Ref. [29] shows that participation in clean-up activities strengthened students' insight into the impact of plastic pollution on the environment. Several studies providing further insight have been conducted in the UAE, where, in Ref. [31], it was shown that students at the secondary school level possess knowledge and awareness about the impact of plastic pollution on the environment and health, as the findings indicated that the students regarded plastic waste as being harmful to the environment. Ref. [32] describes a study conducted in Denmark where students participated in the clean-up of marine waste through Citizen Science (57,000 students, 7–16 years). The study found that students were generally very concerned about plastic pollution, which they explained was due to their increased awareness about the subject. In Italy, a four-day Sea Cleaner programme was conducted that boosted students' attitudes towards and understanding of marine littering [28]. Another similar programme was undertaken in Spain, Italy, Cyprus and Portugal [33] in order to raise awareness about marine littering in the Mediterranean and the Atlantic Ocean. An Indonesian study showed that teaching had a positive but limited impact on students' awareness of marine littering [34].

In this case study, we wanted to investigate whether or not a modelled sustainability project, including field and laboratory work, changed Norwegian secondary school students' knowledge and understanding of how biodiversity is affected by marine plastic pollution. The aim was also to identify what academic expertise and practical activities should be integrated into lessons about marine plastic pollution. This led to the following research questions:

1. Does field and laboratory work in marine plastic pollution education affect secondary school students' knowledge and understanding about biodiversity?

2. What factors influence secondary school students' learning outcomes during practical work in their marine plastic pollution education?

3. Methods

3.1 Research context

During the autumn of 2021, 50 secondary school students in 10th grade (age 15) from a district school in Trøndelag, Norway, undertook a sustainability project on the subject of marine plastic pollution. During the education programme, the students studied the properties of plastic and learned about its impact on biodiversity during field and laboratory work. Field and laboratory work was carried out at Mausund and at the Mausund Field Station (Appendix B, **Table B1**). There, they saw and experienced real-life issues relating to marine plastic pollution. The Mausund Field Station is working towards achieving sustainable development in the ocean, and the area was well suited for the theme of the sustainability project because large quantities of marine waste are flushed ashore on the islands due to the ocean currents. Since 2017, the coastal renovators employed at the Mausund

Field Station have cleaned up marine plastic waste on the coast of central Norway. Mausund' s topography resembled the students' local environment, and it was therefore important to facilitate activities that the students experienced as being relevant and meaningful.

In August 2021, the research group went to the students' secondary school and informed them about the excursion area and what they would be doing there. The pre-test was conducted in order to identify the students' prior knowledge about the subject. After answering the pre-test, they attended an academic lecture on the history, build-up, characteristics of plastics in the ocean, biodiversity and why plastics and microplastics pose a threat to biodiversity, as well as economic, moral, biological and aesthetic arguments (Appendix B, **Table B2**). The students had the opportunity to discuss in plenary the pros and cons of plastic. The researchers led the discussion and gave the secondary school students answers and follow-up questions.

In October 2021, the selected 50 secondary school students were divided into two groups. One group of 25 secondary school students travelled to the excursion area and undertook the education programme on day 1. The second group of 25 secondary school students travelled to the excursion area and undertook the same education programme on day 2.

In the excursion area at the Mausund Field Station, the secondary school students were able to study plastic both in the field and at the field station. The student groups consisting of 25 students were divided into field and laboratory groups, with 12–13 students in each group. Three hours were allocated for fieldwork, and three hours for laboratory work, for each group (Appendix B, **Table B1**).

3.2 Research approach

This study is a descriptive case study and is a strategy that, in qualitative research, describes what is characteristic within a group in a real-world context and where the case is time-sensitive [35]. Descriptive studies are particularly useful for obtaining information relating to the current status of activities in order to provide answers to questions such as what, when, where and how [36]. Both quantitative and qualitative methods were used for answering the research questions. Data was collected in pre-test and post-test questionnaires and group interviews.

The quantitative questionnaire was designed to assess secondary school students' (i) knowledge about how and why one should take care of biodiversity, (ii) knowledge about microplastics, (iii) knowledge about sustainable development seen from a global perspective and (iv) whether or not student actions make a difference. In order to assess the secondary school students' learning outcomes after the laboratory work, the post-test also included (v) knowledge obtained from the laboratory work and (vi) motivation for biodiversity conservation. The questionnaire was designed on the basis of the interdisciplinary subject entitled "Sustainable Development"; the core element entitled "The Earth and Life on Earth" and the science competence goal for 10th grade entitled "Provide examples of and discuss current dilemmas relating to the exploitation of natural resources and the loss of biodiversity" [37]. The questionnaire consisted of open (O) and closed (C) questions (**Table 1**). When the informants were supposed to answer either yes or no, they were asked to justify their answers. Questions 1, 2, and 3 about biodiversity and microplastics were open, question 4 about microplastics was closed, and questions 5 and 9 about sustainable development and the belief that action makes a difference were open. In the post-test, question 6 about laboratory work was open, and questions 7 and 8 about knowledge and understanding after the laboratory work, and about motivation to preserve biodiversity, were closed.

The interview guidelines for the semi-structured interview (Appendix A, **Table A1**) were designed with introductory questions about what the laboratory work day had been like. The questions were formulated around practical work in the laboratory, sustainable development, biodiversity, plastics, the three tests students had conducted in the laboratory, and what the students believed could be done to prevent further marine plastic pollution. To validate the questions and ensure clear and unambiguous questioning, a pilot test was conducted where two colleagues at the same university answered and tested the questionnaire. The feedback resulted in minor changes being made to the wording.

3.3 Data collection instrument

Data was collected during the autumn of 2021 *via* an anonymous pre-test and post-test questionnaire designed on the University of Oslo (UiO)'s digital online form and two group interviews. The research project was conducted in accordance with the regulations on the processing of personal data [38]. The audio recordings from the interviews were deleted at the end of the project.

Of the 50 secondary school students who were asked to participate, 46 gave their consent and answered the pre-test (August 2021), and 37 secondary school students answered the post-test (October 2021). The post-test was carried out at the school on the day after the secondary school students had undertaken the field and laboratory work at Mausund. A group of volunteers from among the 50 secondary school students attended the group interviews immediately after the laboratory work. The group interviews with a total of nine secondary school students were conducted in October 2021. There was an even distribution between the genders, with five boys and four girls. Five secondary school students attended the group interview on day 1, and four secondary school students attended on day 2.

3.4 Data analysis

The data analysis had a qualitative and quantitative approach. The responses and rationale provided by the informants with respect to each questionnaire question were converted into numerical form (score), thus making it easier to summarise the data [36]. The scores were 1 = do not know, 2 = below average, 3 = average, 4 = above average, 5 = very well answered. How well the answers were explained and argued formed the basis for where they were categorised in scores 2–5. The results for each score were provided as a frequency (%) in order to compare whether or not there were changes in the pre-tests and post-tests. Excerpts from the secondary school students' responses to the pre-test and post-test questionnaires and direct quotes and excerpts from the transcribed interviews provide in-depth information with respect to the questionnaire's quantitative analysis.

4. Results

Table 1 lists questions 1-9 from the questionnaire, along with the corresponding results (5) for each score.

DOI: http://dx.doi.org/10.5772/intechopen.1009623

		Pre-test					Post-test				
Questions	Score	1	2	3	4	5	1	2	3	4	5
1. Biodiversity is the diversity of living organisms. In what ways can we take care of biodiversity? (O)		78.0%	13.0%	8.7%	0%	0%	28%	8.1%	51.4%	5.4%	0%
2. Why should we take care of biodiversity? Explain your answer. (O)		60.9%	21.7%	15.2%	2.2%	0%	40.5%	29.7%	21.6%	8.1%	0%
3. Do you know what microplastics are? What are they and where do they come from? (O)		19.6%	21.7%	58.7%	0%	0%	13.5%	0%	35.1%	51.4%	0%
4. Do microplastics pose a threat to biodiversity? Explain your answer. C:		32.6%	32.6%	34.8%	0%	0%	24.3%	16.2%	29.7%	18.9%	10.8%
5. In order to create sustainable development, the international community must work together in three areas. Which ones? (O)		43.5%	30.4%	19.6%	6.5%	0%	56.8%	16.2%	13.5%	13.5%	0%
6. What do you remember best from the laboratory at the Field Station? (O)							18.9%	48.6%	24.4%	8.1%	0%
7. Have you learnt anything from what we did in the laboratory? C:							24.3%	21.6%	43.3%	10.8%	0%
8. Have you been motivated to preserve biodiversity after your laboratory activities? If so, how have you been motivated? C:							40.5%	21.6%	35.2%	2.7%	0%
9. To what extent do you think your/their actions make a difference? (O).		2.2%	19.69%	60.9%	17.4%	0%	8.1%	21.6	35.1%	35.1%	0%

Table 1.
Details about questions 1-9 in the questionnaire with information on whether or not they were open-ended (O) or closed (C) questions. Pre-test and post-test are divided according to the scores (1-5). The frequency for each score is provided as a %. Score 1: don't know, score 2: below averagely well answered, score 3: averagely well answered, score 4: above averagely well answered, score 5: very well answered.

Question 1 had the highest response rate for score 1 for the pre-test. For the post-test there was a marked increase in score 3 (averagely well answered) of 51.4% and score 4 (above averagely well answered) of 5.4%. In response to question 1 one informant wrote: Dispose of the plastic in the right place, pick up plastic that has gone astray and reduce unnecessary plastic production. Another wrote: we can be more aware, pick up more rubbish and try to make a change. During the group interview one informant gave the following answer:

> *When I dug in the soil there was a piece of rope, the sort that consists of tiny pieces of rope, and they had somehow dissolved inside the soil, so the clump of earth only consisted of pieces of rope like that with a bit of soil. And this made me see how much there actually is.*

During the same interview another informant said: "The plastic is not just what we see, but there's also a lot we do not see." A third informant added: "Microplastics. They are everywhere. We looked at a small clump under the microscope and there was a lot!" Another informant explained:

"That experiment which involved studying how much plastic there was in the soil was really surprising."

Question 2 score 1 had 60.9% for the pre-test and 40.5% for the post-test. This showed a change for score 3 (averagely well answered) from 15.2% for the pre-test to 21.6% for the post-test and for score 4 (above averagely well answered) a change from 2.2% for the pre-test to 8.1% for the post-test. A written response to the post-test was: and many animals ingest environmental toxins because the microplastics take it to them. When they ingest environmental toxins, their immune and hormone systems are compromised, and their fertility is reduced.

During the group interview, one of the informants gave the following answer:

> *There was so much plastic in the mink. Yes, I did not actually think there was going to be so much plastic in it since I did not think it ate anything that might resemble plastic or something that had so much plastic in it, so I did not think it [the mink] was going to have as much plastic in it as it did.*

Question 3, score 4 (above averagely well answered) had 0% for the pre-test and 51.4% for the post-test. One informant wrote: There are plastic pieces that are smaller than 5 mm. These come from larger pieces of plastic that have been in nature for a long time.

Question 4 had 0% for both scores 4 (above averagely well answered) and 5 (very well answered) for the pre-test, 18.9% for score 4 and 10.8% for score 5 for the post-test. One informant with a score of 5 wrote: Yes, it does. Microplastics are eaten by animals and then the environmental toxins pass into the animals. They also get a false feeling of being full up. Another informant wrote: There are major consequences for animals such as polar bears, e.g. having two genders as a result of the environmental toxins contained in microplastics.

Question 5, score 3 (averagely well answered) was 19.6% and score 4 (above averagely well answered) was 6.5% for the pre-test, and for the post-test, this was 13.5% for both scores 3 and 4. One informant indirectly responded to the question about how to create sustainable development when they observed microplastics in the stomach of the mink: "We need to do something to stop this."

Questions 6 and 7 provided information about knowledge and understanding obtained from the laboratory experiments at the field station. For question 6, the post-test result was 48.6% for score 2, 24.4% for score 3 and 8.1% for score 4. One of the informants wrote that there was a lot of plastic in both the soil samples and the mink. For question 7, the post-test result was 21.6% for score 2, 43.3% for score 3 and 10.8% for score 4. One of the informants wrote about the knowledge and skills required for identifying different types of plastic.

For question 8, the post-test result was 21.6% for score 2, 35.2% for score 3 and 2.7% for score 4. One informant wrote: When I saw what the mink had in it and that we need to do something to stop this. During the interview one informant also responded as follows: "I was motivated because the teacher [supervisor from the research group] was enthusiastic."

For question 9, the pre-test result was 19.7% for score 2, 60.9% for score 3 and 17.4% for score 4, and for the post-test the result was 21.6% for score 2 and 35.1% for scores 3 and 4. During the interview the informants said something needs to be done. "We need to pollute less, plastic in nature needs to be cleaned up, less plastic needs to be produced, and this will make a difference."

It emerged from the interviews that secondary school students have heard about Plastic in the Ocean in the media, and they have been working on it at school. One of the informants said during the interview: "It's mainly at school. We wrote a text about it in 8th or 9th grade, and we learnt a lot about it then. And then we saw a lot of news broadcasts about it too." Another informant added: "Yes, there's a lot in the media, these subjects are also being addressed more at school than before and it's more relevant now." One of the informants replied: "Media and natural sciences. We're talking about it at home too, but not as much as we hear at school." Another informant added: "At the moment it's everywhere, there is an increasing amount of attention being paid to the environment around the world, so we hear a lot about it, at least young like us people do."

5. Discussion

The discussion is presented on the basis of the research questions.

5.1 Does field and laboratory work in marine plastic pollution education affect secondary school students' knowledge and understanding about biodiversity?

The consequences of marine plastic pollution can be difficult to understand, just like climate change, if they are only talked about. Climate change and marine plastic littering are two sides of the same issue [39]. The global challenges lead to the degradation of the environment and a loss of biodiversity because species face greater challenges in adapting to the environment. The binding global agreement on plastic pollution contains main areas for measures designed to reduce climate change, pollution and biodiversity loss [40]. One possible analogy for explaining people's relationship with marine plastic pollution and its hidden impact on biodiversity, is climate change. Global warming is well documented scientifically, and human behaviour is widely accepted as being a contributory factor [41, 42]. However, studies show that many people do not actively try to counteract global warming, something that may be associated with there being a psychological distance between theoretical understanding and noticeable changes [43]. In Refs. [44, 45], this is described as a form of paralysis, whereby people understand the problem intellectually but disassociate themselves from it emotionally (splitting), and in Ref. [46], indifference is explained as being an emotional response to a constant flood of overwhelming environmental news, which also makes it difficult for many people to see the direct link between human activities and global changes [47].

The results of this study show that the students acquired more knowledge and understanding about microplastics and their impact on biodiversity. The results of the post-test with respect to question 1 showed 51.4% for score 3, compared to 8.7% for the pre-test. This is the knowledge and understanding that is most likely acquired during lessons because score 1 was 78.0% in the pre-test (**Table 1**). This was related specifically to how much microplastic they detected in the soil and in the stomach of the mink during the laboratory work carried out at the field station and what they experienced when they uncovered marine waste in the soil during their fieldwork. When they studied the stomach contents of the mink after dissecting it, the secondary school students acquired knowledge and understanding that showed that microplastics are absorbed into the food chain. One important piece of knowledge that they acquired was that microplastics contain environmental toxins that contribute towards disrupting hormonal balance and reproduction and that microplastics are a threat to biodiversity because they also create a false feeling of being full up. The results show that the secondary school students had an increased knowledge and understanding of questions 3 and 4. Here, the results were 51.4% for score 4 in the post-test compared to 0% in the pre-test for question 3, 18.9% for score 4 and 10.8% for score 5 compared to 0% for scores 4 and 5 in the pre-test for question 4 (**Table 1**). One explanation as to why they were more precise in their descriptions of these may be that they remembered a lot from the lecture that was first held at school and then at the field station about microplastics, environmental toxins and hormones.

The theme of microplastics ran like a connecting thread through the teaching programme at the field station [48]. For question 5, there was a change in score 4 from 6.5% for the pre-test to 13.5% for the post-test (**Table 1**). The three areas that require cooperation on sustainable development were addressed in the lectures but not specifically during the field and laboratory work. During the lecture at the school, the secondary school students gained an overview of the many factors that influence social and economic conditions that are required in order to solve environmental problems. Despite this, it is not always easy to repeat this if they do not have ownership of the content. It should be explicitly mentioned, for example, when secondary school students are out cleaning up marine plastic waste or are standing in the laboratory discussing the properties and uses of plastic. In order to acquire control of what is in nature, it is necessary to focus on the economic situation and the workers in order to do the job. For question 9, there was a change in score 4 from 17.4% in the pre-test to 35.1% in the post-test. The secondary school students show that they have an insight into what should be done. In the interviews, it emerged that secondary school students realise that we need to pollute less, that plastics in nature need to be cleaned up and that less plastic needs to be produced. Here, they are indirectly involved in key factors in ESD, such as how overall agreements, strategies and measures can contribute to solutions, without mentioning such explicitly. The secondary school students do not mention how we can reduce pollution or who is responsible for cleaning up plastics from nature. They point out that it needs to be done. Ref. [25] shows that education has the expectations and potential for focussing on hope when it comes to dealing with climate change. This should have a transfer value in education which could also apply to environmental challenges such as marine plastic pollution, where students should have the hope that commitment and actions will make a difference. Commitment is positively related to hope [24].

5.2 What factors influence secondary school students' learning outcomes during practical work in their marine plastic pollution education?

Secondary school students derive much of their knowledge from media and can repeat information about plastic pollution, but often without a deeper understanding or the ability to justify their claims [31]. The results obtained in the interviews in this study support the fact that secondary school students receive information from the media. The results from the pre-test show that the majority of them have averagely good justifications (score 3) with respect to questions 3, 4, and 9 (58.7, 34.8 and 60.9%) (**Table 1**). Their approach to the complex picture presented by marine plastic pollution must be addressed in several contexts and worked on in several ways.

The secondary school students have acquired increased knowledge and understanding and there may have been many factors that influenced their learning process. It is therefore difficult to determine if it was solely the field and laboratory work that led to their increased learning outcomes. During their interview, one of the informants said that the teacher's [the research group's supervisor] commitment motivated learning, thus creating a greater understanding of the challenge of marine plastic littering. This is consistent with what has been underlined in [21], that is, the teacher's lessons about marine littering are an important factor with respect to the teaching benefits.

Other factors that could influence motivation for learning include teaching being conducted at the Mausund excursion area, which is a different arena than the students' school area. This may have affected students at an affective level because the field and laboratory work was carried out in a local environment [49], and the impressions that they acquired from their relevant experience contributed towards motivation for learning. Reference [50] refers to a connection between external motivation and concerned reactions, thus indicating a high affective level because looking after the environment entails an awareness of the importance of action in order to reduce specific environmental challenges.

The secondary school students' visit was guided by others who were not their usual teachers. The questions that the informants answered were directly related to the laboratory work. They also carried out fieldwork on cleaning up marine waste and identifying its origins on the same day, and this would also affect their overall impressions, commitment and motivation. The teaching was facilitated so that the students could be involved in authentic issues and solutions designed to promote knowledge development and a sense of responsibility [48], and when secondary school students are taught to face current real-world problems, it is more likely that they will be able to solve the problems that they will face in the future [51]. In [21], emphasis is placed on the teacher's role in teaching about marine plastic pollution, where practical ways of working are highlighted as being useful tools for providing students with specific examples and engaging activities.

Schools shall provide students with the skills required for looking after the environment and making sustainable choices. Despite the increase in focus on sustainability both in educational policy and in a research context, there are several teachers who feel uncertain about the academic content and the didactic approach to the subject of sustainability [52]. The background for this can be linked to the fact that sustainability subjects not only provide a theoretical introduction to the subject, but that teaching that introduces the subject in the classroom shall also provide students with sustainability awareness and the competence to act. This is a challenging but important mission in order to ensure that students have an all-around general natural sciences education. The link between ESD and natural science practice has great potential for the development of students' knowledge and skills both inside and outside the classroom [53]. A lot of what has been published about ESD includes the practitioners' gut feelings and political recommendations [10], and there are calls for what is being implemented and the learning outcomes that are emerging from such.

In Ref. [54], reference was made to the report entitled "Education for Life and Work" and the fact that competence transfer of skills and knowledge will be sought after in the twenty-first century, and these should be competencies that students acquire experience of during their education. Ref. [20] refers to the importance of sustainability competencies, where practical work in particular can help students to develop these competencies, as well as equip them with the skills that they need in order to be able to handle complex issues.

6. Conclusion

This study was conducted with 10th-grade Norwegian secondary school students, and the new thing about this study is that it investigated how a modelled sustainability project could contribute to increased knowledge and understanding about how marine plastic pollution affects biodiversity. The secondary school students responded with words and terms that they had been taught in lectures. They referred to the subjects discussed during the lectures and laboratory exercises, and their fieldwork experiences, and this helped the secondary school students to formulate evidence-based arguments about questions on how microplastics affect biodiversity.

The post-test showed a result of 51.4% for score 3 (averagely well answered) and 5.4% for score 4 (above averagely well answered), compared with 8.7% and 0% for the pre-test on how to safeguard biodiversity. The results of the post-test on microplastics were 51.4% for score 4 (above averagely well answered) and 0% for the pre-test. When asked why microplastics pose a threat to biodiversity, the results were 18.9% for score 4 (above averagely well answered) and 10.8% for score 5 (very well answered), and 0% for scores 4 and 5 in the pre-test. These results can be explained by the fact that microplastics were a consistent topic at the field station and that the secondary school students were allowed to study the characteristics and structure of microplastics during two out of three parts of their laboratory work.

Concerned reactions, looking after the environment and the belief that action will make a difference and provide hope for the future may have helped to influence the affective level of the secondary school students and their motivation for learning. Secondary school students realised that we need to pollute less, that plastics in nature must be cleaned up and that less plastic must be produced. When asked if they thought that their actions make a difference, the result was 35.1% for score 4 (above averagely well answered) for the post-test. The findings may reflect secondary school students' personal experiences of the impact of marine plastic pollution in their local environment.

6.1 Future research directions

This study highlights the importance of marine plastic pollution education and opens several pathways for further research. Conducted with 10th-grade students, the teaching program demonstrated its potential for adaption across different academic levels, offering valuable insights for future educational strategies on this vital issue of biodiversity loss. Future research may examine, among other things, how this teaching programme affects students' knowledge and action competence over time, how commitment affects hopes for preserving biodiversity, and whether or not the same student group or other student groups will use these sustainability competencies later on in life. This could include studies in sixth form college (e.g. with respect to subjects like biology and chemistry), or at a higher university level and in teacher education. Furthermore, it might also be interesting to explore how such a teaching programme could be integrated into interdisciplinary projects that include the three sustainability dimensions: economics, social conditions and the environment.

Acknowledgements

Thanks to Eider AS Mausund Field Station for facilitating the field work and for providing guidance for the secondary school students during the completion of the study.

Appendix A

Subject:	Question
Practical work in the laboratory	What was it like doing something practical in the laboratory? What are your science lessons usually like? Do you learn best from doing things in a practical way or from normal whiteboard teaching? Do you feel like you learn more from practical work? Do you feel more motivated by doing practical work? (Outer/inner) What are you motivated by?
Sustainable development	What did you know previously about marine plastic pollution? Where have you heard about it before? Media, school, friends, family, own observations of your local environment? Where do you think your knowledge on the subject comes from? Is it from natural science subjects? Have you acquired a new insight into the problems relating to plastics after spending a day here at Mausund? In what way?
Biodiversity	Can anyone explain what it means when we say that nature is biodiverse? *Total diversity is seen at three different levels: (genetic, species and ecosystems)* What are the arguments for preserving biodiversity? "Do you think it's important to preserve biodiversity, why?" What do you think is the most important reason for maintaining biodiversity? "After today have, you been motivated to preserve biodiversity?"
Plastics	Why do plastics (microplastics) pose a threat to biodiversity? What are microplastics? What did you know about microplastics before and what do you know now? Did you learn anything new?
Laboratory	During your laboratory activities (Items 1–3), what do you find most enjoyable? Did you experience anything that you did not expect in advance? Have you learnt anything that you would like to take with you into your everyday life, and if so, what?
Conclusion	What three things do you think are the most important things we have talked about?

Table A1.
Interview guide.

Appendix B

The laboratory booklet contains information about	
1	Why the area around the Mausund Field Station is eminently suitable for conducting fieldwork on the subject of marine plastic pollution.
2	Facts about plastic
	• decomposition time • effects on biodiversity • affinity of environmental toxins for microplastics • statistics on how many tonnes of plastic end up in the ocean each year • illustrative images
3	Posts at the Field Station with procedures for each laboratory experiment
	Post 1: Investigating soil samples for microplastics Post 2: Identification of plastics (six attempts) in order to identify different types of plastic according to specific criteria. Post 3: Dissection of mink. The stomach of the mink was to be opened in order to check whether it contained microplastics.
4	Field for notes to write hypotheses and results
5	Discussion questions for each laboratory experiment

Table B1.
Information in the laboratory booklet.

Lecture at school before excursion	Lecture at the field station
History of plastics Build-up of plastics Properties of plastic Plastics in the ocean BD* Arguments about why we should look after BD Why plastics pose a threat to BD in the ocean. Microplastics Why microplastics pose a threat to BD Economic, moral, biological, aesthetic arguments. Norway's plastics strategy	Build-up of plastics Properties of plastic Plastics in the ocean BD Arguments on why we should look after BD Why plastics pose a threat to BD in the ocean Microplastics Why microplastics pose a threat to biodiversity

**BD = Biodiversity.*

Table B2.
Academic content of the lectures.

Author details

Viktoria Smenes[1] and Hilde Ervik[2]*

1 Vika Upper Secondary School, Oslo, Norway

2 The Norwegian University of Science and Technology (NTNU), Trondheim, Norway

*Address all correspondence to: hilde.ervik@ntnu.no

References

[1] Brundtland GH. Global change and our common future. Environment: Science and policy for sustainable development. Environment: Science and Policy for Sustainable Development. 1989;**31**(5):16-43

[2] European Commission: Joint Research Centre and Bianchi G. Sustainability Competences – A Systematic Literature Review. Publications Office; 2020. Available from: https://data.europa.eu/doi/10.2760/200956

[3] Leicht A, Combes B, Byun W, Agbedahin A. From agenda 21 to target 4.7: The development of education for sustainable development. In: Leicht A, Heiss J, Byun W, editors. Issues and Trends in Education for Sustainable Development. Paris: UNESCO; 2018. pp. 25-38

[4] Mulà I, Cebrián G, Junyent M. Lessons learned and future research directions in educating for sustainability competencies. In: Vare P, Lausselet N, Rieckmann M, editors. Competences in Education for Sustainable Deveopment. Switzerland: Springer Nature; 2022

[5] UNESCO. Education for Sustainable Development Goals. Learning Objectives. UNESCO; 2017. Available from: https://www.unesco.org/en/articles/education-sustainable-development-goals-learning-objectives

[6] Barth M, Godemann J, Rieckmann M, Stoltenberg U. Developing key competencies for sustainable development in higher education. International Journal of Sustainability in Higher Education. 2007;**8**(4):416-430

[7] Wiek A, Withycombe L, Redman CL. Key competencies in sustainability: A reference framework for academic program development. Sustainability Science. 2011;**6**(2):203-218

[8] Redman A, Wiek A. Competencies for advancing transformations towards sustainability. Frontiers in Education. 2021;**6**(484):1

[9] Wiek A, Redman A. What do key competencies in sustainability offer and how to use them. In: Vare P, Lausselet N, Rieckmann M, editors. Competences in Education for Sustainable Development. Switzerland: Springer Nature; 2022

[10] Boeve-de Pauw J, Gericke N, Olsson D, Berglund T. The effectiveness of education for sustainable development. Sustainability. 2015;7:15693-15717

[11] Olsson D, Gericke N, Boeve-de PJ. The effectiveness of education for sustainable development revisited – a longitudinal study on secondary students' action competence for sustainability. Environmental Education Research. 2022;**28**(3):405-429

[12] Torsdottir AE, Olsson D, Sinnes AT. Developing action competence for sustainability – Do school experiences in influencing society matter? Global Environmental Change. 2024;**86**:1

[13] UNEP. Plastic Pollution & Marine Litter. UNEP; 2024. Available from: https://www.unep.org/topics/ocean-seas-and-coasts/ecosystem-degradation-pollution/plastic-pollution-marine-litter

[14] Sutherland W, Clout M, Côté I, Daszak P, Depledge M, Fellman L, et al. A horizon scan of global conservation issues for 2010. Trends in Ecology & Evolution. 2010;**25**:1-7

[15] Tekman MB, Walther BA, Peter C, Gutow L, Bergmann M. Impacts of Plastic Pollution in the Oceans on Marine Species, Biodiversity and Ecosystems. Germany, Berlin: WWF; 2022

[16] Bergmann M, Collard F, Fabres J, Gabrielsen GW, Provencher JF, Rochman CM, et al. Plastic pollution in the Arctic. Nature Reviews Earth & Environment. 2022;**3**(5):323-337

[17] Balali-Mood M, Naseri K, Tahergorabi Z, Khazdair MR, Sadeghi M. Toxic mechanisms of five heavy metals: Mercury, lead, chromium, cadmium, and arsenic. Frontiers in Pharmacology. 2021;**12**:643972

[18] Ford HV, Jones NH, Davies AJ, Godley BJ, Jambeck JR, Napper IE, et al. The fundamental links between climate change and marine plastic pollution. Science of the Total Environment. 2022;**806**:8

[19] Wagner M. Solutions to plastic pollution: A conceptual framework to tackle a wicked problem. In: Bank MS, editor. Microplastic in the Environment: Pattern and Process. Environmental Contamination Remediation and Management. Cham: Springer; 2022. DOI: 10.1007/978-3-030-78627-4_11. Available from: https://link.springer.com/chapter/10.1007/978-3-030-78627-4_11

[20] Risopoulos-Pichler F, Daghofer F, Steiner G. Competences for solving complex problems: A cross-sectional survey on higher education for sustainability learning and transdisciplinarity. Sustainability. 2020;**12**:13

[21] Ahmad-Kamil EI, Syed Zakaria SZ, Othman M. What teachers should know for effective marine litter education: A scoping review. Sustainability. 2022;**14**(7):4308

[22] Hartley BL, Pahl S, Holland M, Alampei I, Veiga JM, Thompson RC. Turning the tide on trash: Empowering European educators and school students to tackle marine litter. Marine Policy. 2018;**96**:227-234

[23] Bettencourt S, Costa S, Caeiro S. Marine litter: A review of educative interventions. Marine Pollution Bulletin. 2021;**168**:14

[24] Ojala M. Hope and climate-change engagement from a psychological perspective. Current Opinion in Psychology. 2023;**49**:3-4

[25] Ott A. Climate change and education in shades of blue: Between darkness and light with agential realism and object-oriented ontology. Environmental Education Research. 2024;**30**(11):2093-2107

[26] Charitou A, Aga-Spyridopoulou RN, Mylona Z, Beck R, McLellan F, Addamo AM. Investigating the knowledge and attitude of the Greek public towards marine plastic pollution and the EU single-use plastics directive. Marine Pollution Bulletin. 2021;**166**(8):1

[27] Davison SMC, White MP, Pahl S, Taylor T, Fielding K, Roberts BR, et al. Public concern about, and desire for research into, the human health effects of marine plastic pollution: Results from a 15-country survey across Europe and Australia. Global Environmental Change. 2021;**69**

[28] Locritani M, Merlinob S, Abbatec M. Assessing the citizen science approach as tool to increase awareness on the marine litter problem. Marine Pollution Bulletin. 2019;**140**:320-329

[29] Torres HR, Reynolds CJ, Lewis A, Muller-Karger F, Alsharif K, Mastenbrook K. Examining youth

perceptions and social contexts of litter to improve marine debris environmental education. Environmental Education Research. 2019;**25**(9):1400-1415

[30] Chow C-F, So W-MW, Cheung T-Y, Yeung S-KD. Plastic waste problem and education for plastic waste management. In: TLW SCK, Yang M, Chow CF, Tse KH, editors. Emerging Practices in Scholarship of Learning and Teaching in a Digital Era. Singapore: Springer; 2017. pp. 125-140

[31] Hammami MBA, Mohammed EQ, Hashem AM, Al-Khafaji MA, Alqahtani F, Alzaabi S, et al. Survey on awareness and attitudes of secondary school students regarding plastic pollution: Implications for environmental education and public health in Sharjah city, UAE. Environmental Science and Pollution Research. 2017;**24**(25):20626-20633

[32] Oturai NG, Pahl S, Syberg K. How can we test plastic pollution perceptions and behavior? A feasibility study with Danish children participating in "the mass experiment". Science of The Total Environment. 2022;**806**(2):1

[33] Gravina T, Iero A, Giraldo J, Ruiz A, Borges I, Cossu C, et al. MALIA: A project to raise awareness on marine litter in the Atlantic and Mediterranean. Rendiconti Online Societa Geologica Italiana. 2019;**49**:33-40

[34] Kusumawati I, Setyowat M, Syakti AD, Fahrudin A. Enhancing millennial awareness towards marine litter through environmental education. In: The 3rd International Symposium on Marine and Fisheries Research (3rd ISMFR). EDP Sciences; 2020

[35] Priya A. Case study methodology of qualitative research: Key attributes and navigating the conundrums in its application. Sociological Bulletin. 2021;**70**(1):94-110

[36] Babbie E. The Practice of Social Research. USA: Cengage Learning Custome Publishing; 2016

[37] Utdanningsdirektoratet. Læreplan i naturfag (NAT01-04): Fastsatt som forskrift. Læreplanverket for Kunnskapsløftet 2020. 2020. Available from: https://www.udir.no/lk20/nat01-04?lang=nob]

[38] Sikt. Notification Form for personal data 2024. Available from: https://sikt.no/en/notification-form-personal-data

[39] Fava M. Ocean Plastic Pollution an Overview: Data and Statistics Ocean Literacy Portal. IOC UNESCO; 2022. Available from: https://oceanliteracy.unesco.org/plastic-pollution-ocean/

[40] Ministry of Climate and Environment. Global avtale mot plastforurensing. 2023. Available from: https://www.regjeringen.no/en/aktuelt/agreement-in-nairobi-on-global-deal-against-plastic-pollution/id2902765/

[41] Goldberg MH, Gustafson A, van der Linden S, Rosenthal SA, Leiserowitz A. Communicating the scientific consensus on climate change: Diverse audiences and effects over time. Environment and Behavior. 2022;**54**(7-8):1133-1165

[42] Lynas M, Houlton B, Perry S. Greater than 99% consensus on human caused climate change in the peer-reviewed scientific literature. Environmental Research Letters. 2021;**16**:1

[43] Petrescu-Mag RM, Burny P, Banatean-Dunea I, Petrescu DC. How climate change science is reflected in people's minds. A cross-country study on people's perceptions of climate change. International Journal of Environmental Research and Public Health. 2022;**19**(7):1

[44] Randall R. Loss and climate change: The cost of parallel narratives. Ecopsychology. 2009;**1**(3):118-129

[45] Lertzman RA. Myth of apathy: Psychoanalytic explorations of environmental degradation. In: Weintrobe S, editor. Engaging with Climate Change. London: Routledge; 2010

[46] Moser SC. More bad news: The risk of neglecting emotional responses to climate change information. In: Moser SC, Dilling L, editors. Creating a Climate for Change: Communicating Climate Change and Facilitating Social Change. Cambridge University Press; 2007. pp. 64-80

[47] Doherty TJ, Clayton S. The psychological impacts of global climate change. American Psychologist. 2011;**66**(4):265

[48] Smenes VA. Undervisning for bærekraftig Utvikling i Havet [Education for Sustainable Development in the Ocean]. Norway: NTNU Open: The Norwegian University of Science and Technology (NTNU); 2022

[49] Gabrielsen A, Korsager M. Nærmiljø som læringsarena i undervisning for bærekraftig utvikling. In: En analyse av lærerens erfaringer og refleksjoner. Nordina: Nordic Studies in Science Education; 2018

[50] Juma-Michilena J-M, Ruiz-Molina M-E, Gil-Saura I, Belda-Miquel S. How to increase students' motivation to engage in university initiatives towards environmental sustainability. Journal of Consumer Affairs. 2023;**57**:1304-1323

[51] Laurie R, Nonoyama-Tarumi Y, McKeown R, Hopkins C. Contributions of education for sustainable development (ESD) to quality education: A synthesis of research. Journal of Education for Sustainable Development. 2016;**10**(2):226-242

[52] Dahl T. Prepared to teach for sustainable development? Student teachers' beliefs in their ability to teach for sustainable development. Sustainability. 2019;**11**(7)

[53] Eilks I. Science education and education for sustainable development–justifications, models, practices and perspectives. Eurasia Journal of Mathematics, Science and Technology Education. 2015;**11**(1):149-158

[54] Pellegrino JW, Hilton ML. Education for life and work: Developing transferable knowledge and skills in the 21st century. In: Council R, editor. National CoDDLasCSCfEDoBaSSaE. Washington, DC: National Academies Press; 2012

Chapter 4

A Review Study of Environmental Science and Education from a Norwegian Archipelago

Hilde Ervik

Abstract

This chapter describes studies from the Froan archipelago in Frøya municipality in Central Norway. This area contains the Froan nature reserve and landscape protection area, which is the largest continuous coastal area protected as a nature reserve in Norway. The islands of the Froan archipelago are polluted with marine plastic, carried by ocean currents from both Norway and abroad. The marine plastic washed ashore ends up in the ground and in small freshwater lakes that characterise the landscape. The studies from 2012 to 2025 cover environmental monitoring of marine species (fish, edible crabs and kelp) and environmental analyses in small freshwater lakes and ponds and in soil. They describe how teaching on marine plastic pollution has been provided to secondary school students and included in the training of preservice teachers and in continuous professional development (CPD) courses for in-service teachers. Knowledge of the environmental challenges posed by marine plastic pollution is necessary if we are to restore nature to its natural state and prevent the pollution from continuing.

Keywords: environmental monitoring, sustainability, fish, edible crab, kelp, marine plastic pollution, experience-based knowledge

1. Introduction

Mausund is the largest fishing village in Frøya municipality in Central Norway. The archipelago off Frøya consists of 5400 islets and skerries and extends from Sula in the northwest to Halten in the northeast of Frøya (**Figure 1**).

Froan nature reserve and landscape protection area and wildlife sanctuary, bordering Mausund, covers an area of 480 km^2 and contains important habitats for e.g., eiders (*Somateria mollissima*), shags (*Phalacrocorax aristotelis*), white-tailed eagles (*Haliaeetus albicilla*), black guillemots (*Cepphus grylle*), kittiwakes (*Rissa tridactyla*), red-breasted mergansers (*Mergus serrator*), otters (*Lutra lutra*) and grey seals (*Halichoerus grypus)* [1].

One of the shallow sections (63.8°–64.2°N, 8.2° – 9.0°E) of the Frohavet Sea area is described as the 'Mausund Bank' and is a shallow bank with complex bathymetry and small islands [2]. The Norwegian Coastal Current (NCC) runs

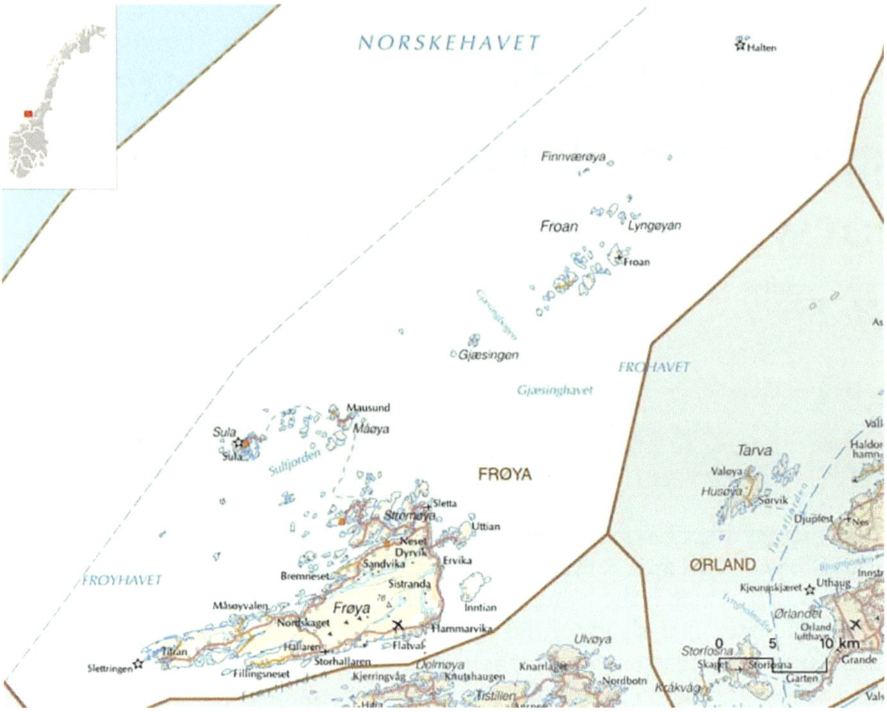

Figure 1.
Frøya municipality. Source: Norwegian mapping authority/Great Norwegian Encyclopedia (Store Norske Leksikon).

north along the coast, above the North Atlantic Current (NAC) [3]. North Atlantic Water (NAW) occasionally washes over the shelf break with the waves from the steep continental shelf, driven by winds and strong local tidal currents, forming an upwelling area with nutrient-rich, saline and warm water around Mausund Bank and Frohavet [2, 4, 5]. This provides the area with good conditions for fishing and aquaculture [6].

Commercial activities at Mausund and in the Froan archipelago are mainly fishing and aquaculture. In the aquaculture industry, Atlantic salmon (*Salmo salar*) predominate, but Atlantic cod (*Gadus morhua*) are also farmed. The cultivation of sugar kelp (*Saccharina latissima*) is a growing maritime industry in Frohavet.

Apart from boat and ship traffic, there are no known industrial pollution sources in this area. Based on this information and the fact that Mausund is located close to the Mausund Bank upwelling area, it was not expected that elevated amounts of pollutants would be found in the marine species examined.

Until a waste collection system was established at Mausund and on the other inhabited islands in the archipelago, much of the household and other waste was dumped in the sea. The same was done with waste from ships and other boats. "Out of sight, out of mind." A lot of waste was also burnt on bonfires. Nobody thought it would be a problem, and we cannot blame others for doing it. That was the culture. The ocean is big, and what ends up in the ocean dissolves away. Or so people thought. Today, there is enough information for most people to change their attitudes to plastic waste being disposed of in nature. Norwegian businesses that live by fishing and aquaculture have introduced measures at different levels to limit the disposal of ropes, fishing equipment and other plastic products, as these pollute the ocean and lead to poor living conditions for the marine species.

Everything that floats and is transported by the NCC and NAC is carried with the waves and washed ashore on the islands, also in the archipelago in Frøya municipality. People who have grown up in Mausund and on the surrounding islands have been used to seeing a lot of plastic waste and other seaborne rubbish that has been washed ashore. Much of it could be reused, but in recent decades, the amount of plastic waste brought in by ocean currents, from both Norway and abroad, has increased. On those islands that are not inhabited, plastic waste has piled up on the ground and in the freshwater lakes and ponds that are characteristic of this landscape. Nature takes its course, and after a while the marine plastic waste is covered with soil and vegetation. The plastic material weathers and it is not suitable for recycling anyway because there is no way of knowing what substances the plastic contains.

The 'Conservation plan for the protected areas in Froan' from 2015 noted that a lot of plastic pollution had been washed tens of metres inland, onto the salt meadows and into protected inlets, and that regular action has to be taken to remove plastic waste from the beach zone on the Froan islands [7]. This marine plastic pollution is an environmental threat to biota. The threat posed by marine plastic pollution is gaining increased attention. There is currently limited data on the impact of marine plastic waste on nature beyond the fact that animals get caught up in fishing equipment and swallow pieces of plastic and polystyrene and that it is aesthetically ugly, but clean-up measures are necessary to protect the fertile and species-rich but vulnerable coastal areas. It is also essential to include teaching about marine plastic pollution as a subject area in schools and higher education.

The aim of this review study is to give an account of research that has been conducted at Mausund and the surrounding area. The work described in this review started in 2012 with environmental monitoring of marine resources, moving on to document marine plastic waste in the ground and contaminants in the freshwater lakes and ponds on the islands, and looking at how the subject of marine plastic pollution is taught to secondary and upper secondary school students and students in higher education. Then there is fishermen's experience-based knowledge from their upbringing and working life as fishermen at Mausund. The fishermen's experience-based knowledge contributes valuable input to scientific research on marine species, and their knowledge has so far been an under-communicated resource.

2. Method

This review is a compilation of nine studies conducted in 2012 in the area around Mausund (63° N, 008° E), in the Froan archipelago with Froan nature reserve and landscape protection area in Frøya municipality in Central Norway, one study conducted with Australian preservice teachers, and one study conducted with in-service teachers. The themes of this review are environmental monitoring of marine resources, marine plastic pollution, marine plastic pollution education, and experience-based knowledge as complementary information to the natural science research.

There are three studies that have documented toxic and essential elements in edible crab, fish and kelp (2012–2019), one study documented concentrations of potentially harmful elements (PHEs) and organic contaminants (OCs) in freshwater, sediments and marine plastic debris (2025), one study documented macroplastic in soil and peat (2021), two articles are about fishermen's experience-based knowledge

Themes	Sample	Year	Reference	Description
Environmental science	Edible crab and fish	2017	https://doi.org/10.1007/s11356-017-1000-4	Concentrations of essential and toxic elements in brown meat in edible crab and fish fillet.
Environmental science	Edible crab and fish	2019	https://hdl.handle.net/11250/2651485	Concentrations of essential and toxic elements in brown meat in edible crab and fish fillet.
Environmental science	Kelp	2019	https://www.boffinaccess.com/journal-water-technology-treatment-methods/the-kelp-laminaria-2-122/jwt-2-122.pdf	Concentrations of essential and toxic elements in kelp's lamina.
Environmental science	Freshwater, sediment, marine plastic debris	2025	Heliyon—Under review	Concentrations of potentially harmful elements (PHEs) and organic contaminants (OCs) in freshwater, freshwater sediment and marine plastic debris.
Environmental science	Soil and peat	2021	https://hdl.handle.net/11250/2777446	Marine macroplastic in soil and peat.
Experience-based knowledge	Fishermen	2022	https://hdl.handle.net/11250/3044513	Norwegian fishermen's experience-based knowledge.
Experience-based knowledge	Fishermen	2023	https://hdl.handle.net/11250/3099656	Norwegian fishermen's experience-based knowledge.
Marine plastic pollution education	Preservice teachers	2025	Journal of Cleaner production—Under review	Preservice teachers and university in collaboration with a company.
Marine plastic pollution education	Secondary school students	2025	IntechOpen	Marine macro- and microplastic. Field and laboratory work with the aim to safeguard biodiversity.
Marine plastic pollution education	Preservice teachers	2024	https://doi.org/10.1016/j.jclepro.2024.143950	A framework for marine plastic litter education based on Australian preservice teachers' knowledge about and attitudes to marine plastic pollution.
Marine plastic pollution education	In-service teachers	2025	Under review	In-service teachers participated in a continuous professional development (CPD) course.

Table 1.
Overview of five environmental science studies, two studies with experience-based knowledge, and four marine plastic pollution education studies, where all studies have Mausund (63° N, 008° E) in Central Norway as a common denominator.

(2022–2023), and four studies are about marine plastic pollution education (2024 and 2025) (**Table 1**). In addition, marine plastic waste was collected from the island Skogsøya near Mausund, by teachers participating in a continuous professional development (CPD) course at the Norwegian University of Science and Technology (NTNU) in April 2019. The collected marine plastic waste was sorted and counted manually.

3. Results and discussion

3.1 Marine species

In the periods 2012–2015 [8] and 2016–2018 [9], essential and toxic elements were analysed in cod (*Gadus morhua*), halibut (*Hippoglossus hippoglossus*) and edible crab (*Cancer pagurus*). There proved to be high concentrations of cadmium (Cd) in the brown meat in edible crabs and rising levels of aluminium (Al), arsenic (As), boron (B), chromium (Cr), copper (Cu), iron (Fe), mercury (Hg), manganese (Mn), nickel (Ni), selenium (Se) and tin (Sn) (**Table 2**). In the period 2015–2017 [10], essential and toxic elements were analysed in kelp (*Laminaria hyperborea*) (**Table 2**). This was in order to assess whether kelp could be used as a bioindicator for environmental monitoring of marine resources.

There was a marked increase in the average concentration of several elements in brown meat in edible crabs from the period 2012–2015 to 2016–2018. B, Cd, Hg, and Se approx. 2-fold higher. Al, As, Cu, Fe and Sn approx. 3-fold higher. Cr and Ni approx. 4-fold higher (**Table 2**). In 2018 [8], it was found that the high levels of toxic elements

Edible crab Brown meat µg g^{-1} d.w.	**2012–2015 n = 66**	**2016–2018 n = 65**	**X fold higher**	***Laminaria hyperborea* µg g^{-1} d.w.**	**2015–2017 n = 48**
Al	1.41	4.76	3.4	Al	1.18
As	44.2	115	2.6	As	61.1
B	2.89	5.62	1.9		
Cd	10.3	21.8	2.1	Cd	0.82
Cr	0.18	0.74	4.1	Cr	0.10
Cu	36.6	113	3.1	Cu	0.74
Fe	96.0	267	2.8	Fe	10.2
Hg	0.12	0.24	2	Hg	0.30
Mn	5.92	8.31	1.4		
Ni	0.79	3.03	3.8	Ni	0.13
Pb	0.16	0.13		Pb	0.04
Se	4.83	7.37	1.5		
Sn	0.003	0.009	3		
Zn	102	94.6		Zn	43.7

Table 2.
*Average concentrations (µg g^{-1} d.w.) in edible crab (*Cancer pagurus*) brown meat of the elements Al, As, B, Cd, Cr, Cu, Fe, Hg, Mn, Ni, Pb, Se, Sn, Zn in the periods 2012–2015 and 2016–2018, and in kelp (*Laminaria hyperborea*) of the elements Al, As, Cd, Cr, Cu, Fe, Hg, Ni, Pb, Zn in the period 2015–2017.*

in brown meat in edible crabs did not come from geogenic sources, but one suggestion was that the toxic elements could be due to long-distance pollution brought to the area on the ocean currents, or that they could be due to local pollution sources. In 2020 [9], it was discussed whether As, Pb and Cd could have been carried along with long-distance atmospheric aerosols [11–13], and that Cr, Ni and Cu were probably from local sources [14]. The findings on essential and toxic elements in *L. hyperborea* provided a snapshot from the period 2015–2015 and need to be followed up with new analyses where samples of lamina are taken at different seasons. The large and species-rich kelp forest in this area provides shelter for fish fry, adult fish and juvenile and young edible crabs.

3.2 Marine plastic pollution

Coastal renovators employed by Eider AS[1] Mausund Feltstasjon began a professional clean-up of marine waste at Mausund and in the Froan nature reserve in 2017. As of the end of 2024, around 1300 tonnes of marine plastic waste has been removed from this area and from several areas in Frøya and Hitra municipalities. In 2025, there are still large amounts of marine plastic waste in this area, remaining in the terrain, in protected inlets and on the shoreline with salt meadows, and in the coming years there will be a continuous annual influx of marine plastic waste from what is already in the ocean and from new releases. During the clean-up work and inspections, large amounts of marine plastic waste were observed under the vegetation and far down into the soil, as well as in the small freshwater lakes and ponds.

3.2.1 Marine plastic pollution in soil

Soil samples were taken from this area in 2020, and it turned out that there was an average of 25% marine plastic waste in the soil [15]. Each soil sample was a 27-litre lump containing differing amounts of non-biological material. The volume of each soil sample was standardised by filling a plastic box holding 27 litres. The soil samples had to be taken by digging them up with a spade. A soil sampler is usually cylindrical with a diameter of approx. 5–10 cm. The marine plastic waste buried in the ground made it impossible to retrieve the samples with a soil sampler. During the fieldwork and collection of the soil samples, rope ends and other residues, scraps of fishing net, plastic pieces (both soft and hard plastic), and microplastics left behind in the soil were observed (**Figures 2** and **3**). The microplastic observed during the fieldwork was visible because the plastic was bright blue in colour. Contamination of the soil with marine plastic waste will have a negative effect on biota. Over time, the plastic waste will also fragment into plastic debris and secondary micro- and nano-plastics. Water running through this soil is likely to carry the plastic particles back into the ocean.

Among other things, the vegetation also helps to split ropes, as the roots push between the longitudinal fibres that the ropes are made of (**Figure 4**). This shows that marine plastic waste fragments into plastic debris, even though it is covered with soil and hidden by vegetation.

3.2.2 Marine plastic pollution in small freshwater lakes and ponds

Marine plastic waste that has been washed ashore on the islands and has ended up in the small freshwater lakes and ponds often stays there, and over time the plastic

---

1 www.eider.no

Figure 2.
Bits of rope and fishing net in soil samples. Photo: Mausund Field Station.

Figure 3.
Bits of rope and fishing net, plastic pieces (soft and hard plastic), and microplastics remaining in the soil. Photo: Hilde Ervik.

becomes thin and brittle under the effects of UV rays, frost, mechanical influences, and saltwater from when it was transported by ocean and coastal currents. The plastic products therefore break down easily into plastic debris, plastic pieces and secondary micro- and nano-plastics. From personal experience during fieldwork in this area, visible fouling by algae on soft plastic and other plastic products was observed, causing the plastic to sink to the bottom of these small freshwater lakes and ponds (**Figure 5**).

Because this area is heavily contaminated with marine plastic waste, it is well suited for studies to assess whether marine plastic waste could be a local pollution source for a range of potentially harmful elements (PHEs) and organic contaminants (OCs). Several of these PHEs and OCs may be added during the production of the plastic, or PHEs and OCs may be transported to the area with aerosols or sea foam, to then be adsorbed into the surface of the plastic after it has ended up in nature as

Figure 4.
Roots growing into rope and breaking up the fibres it is made of. Photo: Hilde Ervik.

Figure 5.
Marine plastic waste in one of the small freshwater lakes. Photo: Hilde Ervik.

marine plastic waste. PHEs and OCs in marine plastic waste can leak out into the surrounding environment, in the water and sediments, or can be absorbed into organisms and vegetation. Analysis of PHEs and OCs in water, sediments and marine plastic waste obtained from these small freshwater lakes and ponds can provide evidence of whether marine plastic waste is a local source of pollution. PHEs and OCs in the small freshwater lakes and ponds have 'nowhere to go' and will therefore become increasingly concentrated over time. There are difficulties in documenting this on beaches and in the ocean because the waves produce constant change and dilution. The knowledge gathered from the small freshwater lakes and ponds may be transferrable to similar issues in ocean and coastal areas.

In 2019 and 2020, samples were taken from a total of 17 small freshwater lakes and ponds on islands in the area from Sula, west of Mausund, to Froan, east of Mausund. In that study, relatively high concentrations of several PHEs and OCs in water and sediments were documented, and the concentrations of several OCs were higher in marine plastic waste than in the sediments [16]. Some OCs have a high biomagnification potential and can accumulate in sheep (*Ovis aries*) and other livestock, wildlife and plants. The OCs that were documented in marine plastic waste and which cause concern are the polyfluoroalkyl substances (PFAS) perfluorooctane sulfonate (PFOS) and perfluorooctanoic acid (PFOA) [16]. Birds and animals swallow marine plastic waste as they think it is food and are then exposed to PFOS and PFOA, among other things. PFOS and PFOA can pose risks to both human health and the environment [17].

3.3 Fishermen's experience-based knowledge

Catches of edible crab and fish were taken in collaboration with an experienced fisherman from Mausund. The fisherman shared his knowledge and expertise during the fieldwork. This included the migration pattern of edible crabs, how fish stocks have changed over time, and how fishing methods have evolved.

The fishermen's experience-based knowledge provides valuable input when findings from fieldwork are to be assessed. Local knowledge of the areas where fish and edible crabs are caught needs to be included in the assessment. In 2022, a study was conducted to document how fishermen have developed their experience-based knowledge [18] and use this to show how their experience-based knowledge can contribute additional knowledge to research-based documentation [19]. One example is the question of what are local pollution sources that could cause toxic elements to increase in brown crab meat, as described in Section 3.1.

One study explained that crabs grow more slowly in colder areas, that toxic elements accumulate in the brown meat over time, and that this combination produces high concentrations of Cd in edible crabs [20]. What edible crabs eat that contains these toxic elements is unclear. The fishermen interviewed in the 2022 study stated that edible crabs move down into the deep water in the winter. There is also video evidence to show that crabs reach depths of more than 400 metres [21]. Fishermen who have set nets to catch angler fish (*Lophius piscatorius*) at depths from 100 to 400 metres in Frohavet have reported that marine plastic waste regularly sticks to the fishing equipment. This indicates that the waste collects in the deep sea trenches of Frohavet and is a local pollution source.

3.4 Marine litter education

In parallel with fieldwork to collect samples documented in the studies referred to in 3.1 and 3.2.2, upper secondary school students, in-service teachers, preservice teachers and secondary school students were taught about marine resources and marine plastic pollution. The emphasis of this teaching varied slightly depending on the target group attending the classes. The following sub-sections give examples of the teaching programmes conducted with upper secondary school students, in-service teachers, preservice teachers and secondary school students. In several of the teaching programmes, the participants contributed 'citizen science' by collecting samples to be included in studies as described in 3.1 and 3.2.2 and elsewhere.

3.4.1 Upper secondary school students

Students in upper secondary school who had the technology and research theory in the curriculum helped to clean up marine plastic waste at Mausund in order to gain an overview of the environmental challenges posed by this waste. They were given introductory talks by the general manager of Eider AS Mausund Feltstasjon. During the field trip to Mausund, the students also worked on project assignments on the subject of environmental monitoring of marine resources and marine plastic pollution. All of the students went out in groups with the local fisherman in his boat and brought in the catch of fish and edible crab; those who chose to look at environmental monitoring of marine resources took samples of fish fillet and brown meat from edible crabs under the supervision of their teacher. The samples had to be processed according to instructions in order to avoid contamination, and each sample was given a separate ID, which was noted on a standard form. The samples were then sent to the laboratory at Norwegian University of Technology and Science (NTNU) to be analysed for essential and toxic elements. Those who chose marine plastic pollution as their project topic examined soil samples and stomach contents from marine species such as mink (*Neovison vison*) and porpoises for microplastics. The students who chose this topic were trained in how to analyse microplastics. The analyses were carried out with a stereo microscope fitted with Night Sea equipment—either a UV filter or a blue filter—with light sources producing either UV or blue light.

3.4.2 In-service teachers

Upper secondary school teachers who attended a continuous professional development (CPD) course at NTNU on the subject of 'Sustainable Development in the Ocean' carried out fieldwork at Mausund. The course started with a talk by the general manager of Eider AS Mausund Feltstasjon, and they were told about the clean-up work on marine plastic pollution on the islands in the area and about the challenges that marine plastic pollution poses. They were taken out to an island and briefed by coastal renovators on where the fieldwork was to be done. During the fieldwork, a 15 x 15 metre square was measured out in a beach area. Ropes were stretched out to mark this area. The teachers, who were science teachers, cleared much of what lay on the ground, as well as what they found under rocks and vegetation, in about two hours. What they cleared was collected in bags and taken back to Mausund Field Station. The area was still not completely cleared of marine plastic waste. At the field station, the teachers sorted the marine plastic waste and counted the number of pieces. The results of this fieldwork are shown in **Table 3** and **Figures 6-9**.

Another year, teachers attended a CPD course at NTNU on the subject of 'Environmental Monitoring of Marine Resources', where they boarded the research vessel Gunnerus in Trondheim and were taken to Mausund. The journey time was about 8 hours, and during the voyage, the teachers heard lectures on topics relating to environmental monitoring of marine resources. These included the marine ecosystem, environmental monitoring, and an insight into previous studies that had documented the environmental status of sediments on the seabed, in fish and in edible crabs. During the fieldwork at Mausund, they were able to observe how the

Marine plastic waste	Rope from the aquaculture industry	Pieces of rope >50 cm.	Pieces of rope <50 cm	A mix of large and small pieces
Quantity	156	578	3257	Approx. 14,500

Table 3.
Marine plastic waste collected in a 15 x 15 metre square plot analysis.

Figure 6.
Rope from the aquaculture industry. Photo: Mausund Field Station.

Figure 7.
Pieces of rope >50 cm. Photo: Mausund Field Station.

crew on board Gunnerus picked up sediment from the seabed using a grab, as well as receiving the catch of edible crab and *L. hyperborea* from the local fisherman. One of the course tutors went with the fisherman to collect *L. hyperborea* and demonstrated how the samples should be taken from the lamina. The teachers took samples of the brown meat from the edible crabs and *L. hyperborea* under the supervision of the tutor. The samples were taken according to instructions to avoid contamination and subsequently sent to the laboratory at NTNU to be analysed for essential and toxic elements.

Many teachers are uncertain about how to conduct teaching programmes on the subject of sustainable development [22], where the focus is often on socio-scientific issues (SSIs). To help more teachers cover marine litter education with their students, a one-day course was given in the science room at NTNU. The teachers were given a model teaching programme that they could later present at their own schools [23]. Soil samples were then taken from islands around Mausund in fish boxes that are

Figure 8.
Pieces of rope <50 cm. Photo: Mausund Field Station.

Figure 9.
A mix of large and small pieces. Photo: Mausund Field Station.

marine plastic waste to study what non-biological material was hidden in the soil samples and examine the soil samples under a stereo microscope (fitted with Night Sea equipment) for microplastics. Dead minks were taken from the same area, which the attendees dissected before examining the stomach contents for plastic debris and microplastics.

3.4.3 Preservice teachers

Preservice teachers who chose 'Sustainable Development in the Ocean' as the subject of their master's thesis, limited to marine plastic litter, worked with the general manager and coastal renovators employed by Eider AS Mausund Feltstasjon during the field trip and fieldwork. This collaboration between the university, the preservice teachers and the company was described in one study [24] as a hybrid learning environment. The preservice teachers also began their stay at Mausund with a talk from the general manager at Mausund Field Station, before going out with the coastal renovators to help in cleaning up marine plastic waste. The preservice teachers then designed a teaching programme for marine litter education, which they themselves presented at Mausund with secondary school students from a neighbouring municipality.

3.4.4 Secondary school students

Students from a secondary school made a field trip to Mausund for two years running, where they received marine litter education from preservice teachers. The day was divided so they could carry out a square plot analysis on a marked 1.5 x 1.5 metre plot in a beach area on an island off Mausund. In the second half of the day, the secondary school students did laboratory work at the field station and examined microplastics in the soil and in the stomach contents of mink, as well as carryied out tests to identify the type of plastic. The tests in the laboratory, where microplastics were a recurring theme throughout the teaching programme, were designed to make the secondary school students more aware of how to safeguard biodiversity where marine plastic pollution is an environmental threat [25, 26].

In a third year, students from a secondary school helped to clean up marine plastic waste and took samples of water, sediments and marine plastic waste from a small freshwater lake/pond on an island off Mausund. They took samples under the supervision of their teachers. They followed instructions during sample collection to avoid contaminating the samples. The samples were sent to a laboratory at NTNU to be analysed for toxic elements and to the Norwegian Institute for Water Research (NIVA) in Oslo for analysis of microplastics in the water samples. The results of these analyses will be included in a new research study.

3.4.5 Pedagogical framework

Based on experience from the planning, which started in 2011 with the design of a training programme at Mausund to teach about environmental monitoring of marine resources and later marine litter education for different target groups, a need has emerged for a guide to what should be covered by marine litter education. Many places in Norway and other countries face similar challenges with marine plastic pollution to those found at Mausund and in the Froan archipelago, with the same need for environmental monitoring of marine resources.

The Great Barrier Reef (GBR) in Queensland, Australia, is a vulnerable area for environmental pollution. In 2023, a study was conducted with Australian preservice teachers to examine their knowledge and attitudes towards marine plastic pollution. It turned out that they were aware of the challenges from marine plastic pollution and expressed concern about the GBR but had less understanding of the impact that chemicals added to or adsorbed into the plastic could have on nature [27]. Findings from this study helped to draft a pedagogical framework for marine litter education [27].

4. Conclusion

In this review study, the environmental challenge of marine plastic pollution is described in the Mausund area in Frøya municipality in Central Norway. The shallow sections of the Frohavet Sea described as Mausund Bank, have a nutrient-rich upwelling area and provide good conditions for fishing and the aquaculture industry. The large and species-rich kelp forest in this area also provides shelter and a good breeding ground for fry, adult fish and juvenile and young edible crabs.

Environmental monitoring of marine species has been ongoing since 2012 in the Mausund area, and as increasing concentrations of several toxic elements are being found, especially in the brown meat in crabs, there is every reason to look more closely at what specific local pollution sources there might be. This review was specifically aimed at marine plastic pollution, as large quantities of marine plastic waste have been documented in this area. There is a need to restore the vulnerable nature areas impacted by marine plastic pollution in the Froan archipelago and the Froan nature reserve and landscape protection area in order to preserve these species-rich ecosystems. Further research and education are needed to take the necessary measures and to prevent the pollution from continuing.

Thanks

Thanks to general manager Odd Arne Arnesen and the coastal renovators at Eider AS Mausund Feltstasjon for their guidance and for arranging and carrying out field and research work at Mausund and on the islands of the Froan archipelago. Thanks to Bjørn Munro Jenssen, Alexandros G. Asimakopoulos, and Geir Johnsen at NTNU for their support to environmental monitoring of marine resources and documenting potentially harmful elements (PHEs) and organic contaminants (OCs). The samples were analysed at the Department of Chemistry at NTNU and contributed to the assessment and presentation of the sample results.

Author details

Hilde Ervik
The Norwegian University of Science and Technology (NTNU), Trondheim, Norway

*Address all correspondence to: hilde.ervik@ntnu.no

References

[1] Halleraker JH. Froan naturreservat og landskapsvernområde Great Norwegian Encyclopedia (Store norske leksikon). Available from: https://snl.no/Froan_naturreservat_og_landskapsvernomr%C3%A5de

[2] Fragoso GM, Davies EJ, Chauton M, Fossum T, Ludvigsen M, et al. Physical controls on phytoplankton size structure, photophysiology and suspended particles in a Norwegian biological hotspot. Progress in Oceanography. 2019;**175**:285

[3] Skagseth Ø, Drinkwater KF, Terrile E. Wind- and buoyancy-induced transport of the Norwegian coastal current in the Barents Sea. Journal of Geophysical Research. 2011;**116**:1

[4] Sætre R. In: Sætre R, editor. Driving Forces in the Norwegian Coastal Current. The Norwegian Coastal Current—Oceanography and Climate. Trondheim: Tapir Academic Press; 2007

[5] Fragoso GM, Johnsen G, Chauton MS, Cottier F, Ellingsen I. Phytoplankton community succession and dynamics using optical approaches. Continental Shelf Research. 2021;**213**:2

[6] Fragoso GM, Dallolio A, Grant S, Garrett JL, Ellingsen I, Johnsen G, et al. The role of rapid changes in weather on phytoplankton spring bloom dynamics from mid-Norway using multiple observational platforms. Journal of Geophysical Research: Oceans. 2024;**129**:2

[7] Bryn A. Skjøtselsplan for verneområdene i Froan. Norsk institutt for bioøkonomi. 2015;**1**(38):30, 36

[8] Ervik H, Finne TE, Jenssen BM. Toxic and essential elements in seafood from Mausund, Norway. Environmental Science and Pollution Research. 2018;**25**:7409-7417

[9] Ervik H, Lierhagen S, Asimakopoulos A. Elemental content of brown crab (*Cancer pagurus*)—Is it safe for human consumption? A recent case study from Mausund, Norway. Science of the Total Environment. 2020;**716**:5-7

[10] Ervik H. The kelp Laminaria hyperborea as a bioindicator. Journal of Water Technology and Treatment Methods. 2019;**2**(1):3-4

[11] Gladney ES, Small JA, Gordon GE, Zoller WH. Composition and size distribution of in-stack particulate material at a coal-fired power plant. Atmospheric Environment (1967). 1976;**10**(12):1071-1077

[12] Greenberg RR, Zoller WH, Gordon GE. Composition and size distribution of particles released in refuse incineration. Environmental Science & Technology. 1978;**12**(5):566-573

[13] Steinnes E, Berg T, Uggerud HT. Three decades of atmospheric metal deposition in Norway as evident from analysis of moss samples. Science of the Total Environment. 2011;**412-413**:351-358

[14] Christensen ER, Steinnes E, Eggen OA. Anthropogenic and geogenic mass input of trace elements to moss and natural surface soil in Norway. Science of the Total Environment. 2018;**613-614**:371-378

[15] Cyvin JB, Ervik H, Kveberg AA, Hellevik C. Macroplastic in soil and peat. A case study from the Remote Islands of Mausund and Froan landscape

conservation area, Norway; implications for coastal cleanups and biodiversity. Science of the Total Environment. 2021;**787**:6

[16] Ervik H, Mészey A, Røsvik A, Jenssen BM. Selected emerging contaminants, PAHs, and toxic elements in water, sediments, and plastic debris, from a coastal archipelago in Norway. Heliyon. 2025:19-20

[17] Wee SY, Aris AZ. Environmental impacts, exposure pathways, and health effects of PFOA and PFOS. Ecotoxicology and Environmental Safety. 2023;**267**:11

[18] Ervik H, Østern TP, Strømme A. A narrative inquiry into fishermen's experience-based knowledge. Journal for Research in Arts and Sports Education. 2022;**6**(4):43-58

[19] Ervik H, Strømme A. Bridging the gap between experience-based knowledge and the scientific knowledge. International Educational Research (IER). 2023;**6**(2):5

[20] Wiech M, Frantzen S, Duinker A, Rasinger JD, Maage A. Cadmium in brown crab *Cancer pagurus*. Effects of location, season, cooking and multiple physiological factors and consequences for food safety. Science of the Total Environment. 2020;**703**:8

[21] Bakke S, Buhl-Mortensen L, Buhl-Mortensen P. Some observations of *Cancer pagurus* Linnaeus, 1758 (Decapoda, Brachyura) in deep water. Crustaceana. 2019;**92**(1):95-105

[22] Dahl T. Prepared to teach for sustainable development? Student teachers' beliefs in their ability to teach for sustainable development. Sustainability. 2019;**11**(7):7

[23] Ervik H. Bærekraftprosjekt–Opplæring om marin plastforurensning (Sustainability project—Marine plastic pollution education). In: Emstad AB, Knutsen B, editors. Utforskende LæringsFormer i Skolen (Exploratory Learning Methods in School): Universitetsforlaget. Oslo, NO; 2025

[24] Ervik H, Rønning B, Remmen KB. Preservice teachers' learning and teaching about marine plastic pollution in collaboration with a company. International Journal of Sustainability in Higher Education. Under review

[25] Smenes VA. Undervisning for bærekraftig Utvikling i Havet (Education for Sustainable Development in the Ocean). NTNU Open: The Norwegian University of Science and Technology (NTNU); 2022. Available from: https://ntnuopen.ntnu.no/ntnu-xmlui/handle/11250/3017057?locale-attribute=en

[26] Smenes VA, Ervik H. Secondary school students` sustainability competences in marine litter education. In: Carmona EC, editor. Island Ecology—The Need for its Research and Education for Conservation. London, UK: IntechOpen; 2025

[27] Ervik H, Evans N, Taylor S. Pre-service teachers' attitudes to and knowledge of marine plastic pollution and its impacts on the natural environment. Journal of Cleaner Production. 2024;**478**:6

Chapter 5

Mapping of Cultural and Natural Heritage Using Advanced Technologies Such as GIS, Drone, Lidar and Laser Scanner: Case Study – Lubonja Village, Korçë, Albania

Edmond Hoxha, Neritan Ceka, Skënder Aliu, Perikli Qirjazi, Jeton Pekmezi, Elio Hobdari, Elva Leka, Emiliano Qose, Erland Alla, Spartak Kuçaj, Era Xhaferaj, Fjoralba Prifti, Genti Proko and Jani Zoica

Abstract

The village of Lubonja is located in the district of Korça in Albania. It has an area of about 10 km^2, a current population of about 800 inhabitants and is positioned at an altitude from 900 m to 1703 m. Lubonja has great cultural and natural heritage values, including 11 potential areas. The study deals with the use of advanced technologies such as Geographical Information Systems (GIS), Drone, Lidar and Laser scanner, for mapping and documenting these heritages. These processes are very important not only for the preservation but especially for the promotion of this wealth and the organization of cultural and natural tourism. The study was focused on potential areas, specifically on the "Castle of Lubonja"; "Dormition of Santa Mary" church from 1015, the centuries-old oaks and pines and the glacial lake of "Mali i Zi". Based on the current data, the study reaches the conclusion that Lubonja is an early settlement that includes the periods Bronx, c. XV-XI BC; Iron, c. XI-VII BC; Roman c. I-IV BC; Late Antiquity, fifth to sixth centuries AD; Modern. "Castle of Lubonja" belongs to the period of Illyrian Antiquity and the Illyrian-Byzantine period, eight to seven centuries BC and five to six centuries AD. The study gives relevant recommendations for the preservation of objects and their scientific, historical, didactic, cultural, religious and touristic values.

Keywords: Lubonja, cultural and natural heritage, Castle of Lubonja, church of "Dormition of Saint Mary", GIS, drone, Lidar, laser scanner

1. Introduction

1.1 Lubonja, general information

The village of Lubonja is located in the South-East of Albania, about 25 km from the city of Korça (**Figure 1**). The relief is very diverse, with inter-mountainous plateaus, quite differentiated and interrupted by the streams that feed the Osumi River. The territory of Lubonja has an altitude of 900-1703 m. The climate is Mediterranean, which

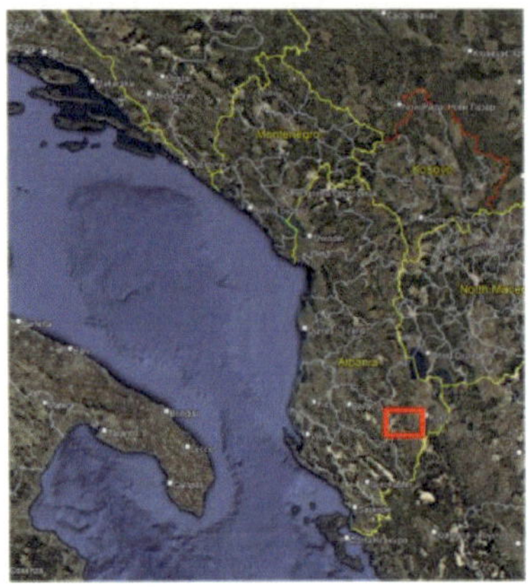

Figure 1.
Geographical position of Lubonja. [Source: Image by Google Earth; Illustration by E.Hoxha].

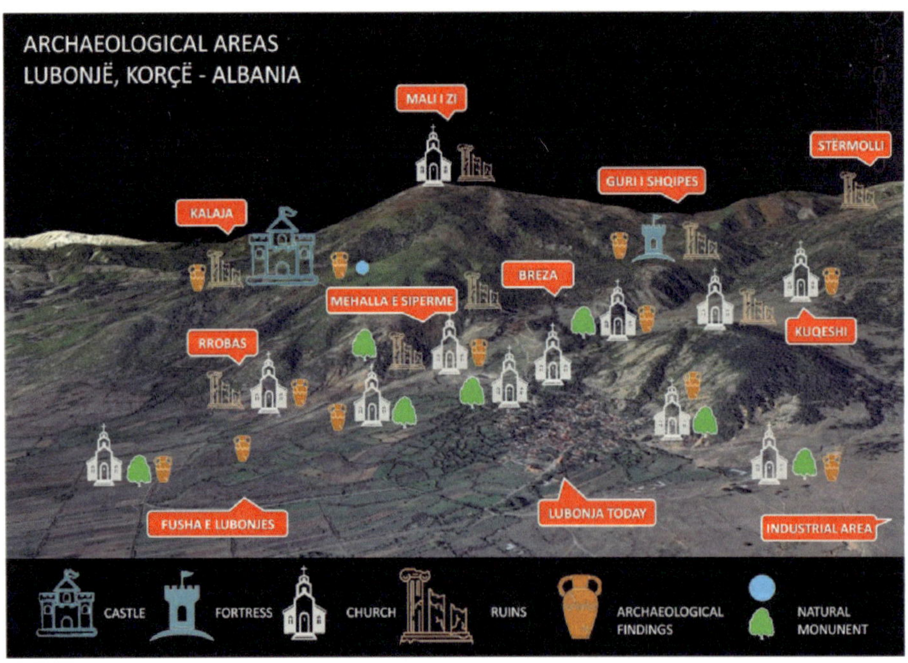

Figure 2.
View from the areas with archaeological potential of Lubonja village. [Source: Image by Google Earth; Illustration by E.Hoxha].

is obviously influenced by the continental climate. The temperature ranges from (+7°) to (−7°), while the maximum (+8) to (+30°). The coldest month is February, while the warmest month is July. The amount of annual rainfall in the area is 710–1050 mm.

1.2 Areas with archaeological potential and natural monuments

The village of Lubonja has 11 areas with archaeological potential: (1) *Current settlement*, Lubonja; where the main objects are cult objects, village bridges and centuries-old oaks. (2) *Kuqeshi*: it is considered as an earlier settlement than the current settlement [1]. (3) *Mëhalla e Sipërme*: located approximately at the same level between Kuqesh and current Lubonja. (4) *Rroba*si: thought to be the earliest settlement and perhaps the beginning of the existence of today's Lubonja [1]; (5) *Prehistoric Castle of Lubonja*: It is located at the elevations of 1200-1240 m. (6) *Mali i Zi*: Located at the highest elevation of 1703 m. (7) *Guri i Shqipes*: It is a special strategic hill. (8) *Breza*: Located about 1300 m above sea level. It is thought that there are ruins and foundations of buildings. (9) *Mountain "Stërmolli"*: It is thought that there were inhabited centers [2]. (10) *Lubonja field*: there are many accidental archaeological finds which make this area very interesting. (11) *Rehova copper mine* industrial area. In terms of natural monuments, Lubonja has evidenced (1) centuries-old oaks near holy places (churches); (2) centuries-old pines near the tower of the Castle of Lubonja and (3) lower Glacial Lake of Mali i Zi (**Figure 2**).

2. Use of GIS, drone, Lidar and laser scanner technologies, in mapping and documenting cultural and natural heritage

2.1 Geographical information systems (GIS)

GIS has become a very important tool in archaeology since the 1990s [3]. Considering that archaeology studies human life in space and time, its combination with GIS makes the process of studying, documenting, filing, processing and visualizing voluminous data very effective, fast and low-cost. What makes the use of GIS in archaeology even more important is not simply the construction of maps, but its capabilities; to merge and analyze different types of data thus creating new information [4]; to store information and assist decision-making; to record, analyze and interpret data directly at the excavation site; and the ability to model and predict in existing areas, but also to discover new areas.

2.2 Drone and Lidar technology

Drone and LiDAR technologies in archaeology are used to discover details and make analyses. One of the most concrete examples is the identification of the ruins of ancient buildings which, due to vegetation and tall trees, are impossible to discover with classical methods. The combination of the LiDAR sensor, the Drone and the Laser Scanner is a valuable aid to reveal details inaccessible to archaeologists. Some of the applications of LiDAR in archaeology are: The study of archaeological remains covered with trees and vegetation; Inventory of cultural heritage; Reconstruction of facades; Construction of virtual 3D models; Separation of archaeological remains from soil and vegetation; Scanning individual parts of archaeological finds, helping to assemble them virtually, before reconstructing them, etc. [5].

2.3 Laser scanner technology

The fact that laser scanning provides a cloud of millions of points with coordinates, representing the surface of the object, or scanned area, allows realizing a polygonal representation of the surface and 3D model of the study area. Laser scanning has many applications in archaeological research and practice: It records the shape, size and condition of sites and objects while minimizing the risk of damage or loss; Capture features that are difficult or impossible to observe or measure by other methods; produces accurate and detailed 3D models that can be used for documentation, analysis, interpretation, visualization and communication of archaeological data and knowledge; Finally, it increases access and preservation of archaeological heritage by creating digital copies that can be stored, shared and displayed online, or in virtual reality. Its combination with geophysical studies, photogrammetry and GIS to create interactive maps and landscapes with multiple layers of archaeological information is very fruitful [6].

3. Archaeological overview of the area of Lubonja: Mapping and documentation of areas with archaeological potential

In general, Lubonja, together with the surrounding villages, located at the end of the southern slopes of Mali i Zi, with an elevation of 1703 m, with rich flora and fauna, with forests and pastures suitable for livestock, with fertile lands for agriculture and viticulture, have favored human life since early in Prehistory. The waters that descend from the slopes of Mali i Zi and from the surrounding hills flow into the Osumi River, which has served as a caravan route since early prehistory. Tracts of stone-paved roads and stone and wooden bridges, ancient and medieval, located during our field surveys, accompanied by archaeological results and Greco-Roman and Byzantine written sources, have enabled us to present the main roads that passed through the territory of Lubonja village. They served to connect with nearby and distant provinces and, at the same time, had a strategic and military character, at the time of the Illyrian-Roman wars and especially during the period of the Avaro-Slavic invasions in the centuries sixth to seventh AD. Convincing in this particular case is the Castle of Lubonja, fortified with prehistoric walls and especially with well-dated walls in late antiquity, when it should have also received the Slavic toponym Lubonja [7]. Several historical roads are evident, but the most important are two. (1) *Adriatic Sea-Aegean Sea Route*. This road started from the shores of the Adriatic, passed along the Osumi valley, through Skrapar, broke off at the end of the Kaltanji's mountain, passed at the end of "Qezari" Castle, Qesaraka, continued over Boshanj, then through the village of Lubonja, where a stone bridge is also preserved, reached Floq. Further down the mountain of Qarri, it reached Castoria (Celetri) and ended in Thessaly and the shores of the Aegean Sea. (2) *Osumi-Korçë river road*. This road is twinned by the Osumi river; at the end of the Kaltanji's mountain, it passed along the Treska river; at the end of Selenica e Pishës, it ended near "Gjoli i Madh", where a tract of the cobbled road is preserved [8], it passed through Lubonja and then ended in the city of Korça.

3.1 Lubonja, current residence

This area includes the churches located inside the village such as "*Dormition of St. Mary*" Church, "*Kamenë*" Church, "*St. George*" Church, "*St. Athanasius*" Church; the old bridge with arches over the village river, the small bridge near the church; and the centuries-old oaks near the churches. From the archaeological point of view, based on the general survey, we can say that taking into consideration the fact that the churches were built in the three dominant places of the village that look at each other, there is a possibility that next to them, or below them, there are mounds. From the quick surface survey, ceramics were found near the "*Kamenë*" Church.

3.2 Kuqeshi

Kuqeshi is located about 3 km to the northwest of the village. It must have been a big village and the name "*Kuqesh,*" according to villagers, is connected to a legend [9]. This area includes (1) "*St. Demetrius*" Monastery; (2) "*St. Friday*" Church; (3) Church of the "*Guri i Madh*"; (4) Area below the drinking water source. During the surveys on the surface, several groups of ceramics were obtained and divided into two groups: Ceramics and Tomb mounds. The first group includes ceramics from two main periods, the fifteenth to fourteenth centuries BC, and Iron from the eleventh to seventh centuries BC. The second group of finds is represented by several hill-grave mounds on the surface of which ceramic fragments were found.

3.2.1 Church or monastery of "Saint Demetrius"

According to legends, this Monastery was built in a field by the road that goes from Lubonja to Kuqesh and further to Stërmolli, Rrungaja and Qafa e Martës, ending to Skrapar. According to stories of residents, this was a very large object, and it is said that there was also a large library [10]. During the survey of the site, ceramics from the iron and bronx periods were found.

3.2.2 "Saint Friday" church or "Hill of tombs above St. Demetrius"

The community is not sure if the "Saint Friday" Church really was here, but based on stories of residents, the presence of many centuries-old oaks; ruins and shapes surrounded by stones, which we think could be graves, it is thought that the Church was here. Not being sure of the name, in this study, we also called it "*Hill of Graves*" above "*Shën Mitri*". From the survey on the surface, the closer observation and the cleaning of a clandestine excavation to a depth of 50 cm, ceramics were found. In addition to finding ceramics, the other reason that makes us think that they are graves is their uniform shape, surrounded by stones and oriented, all in the East–West direction. Such a shape and orientation is reminiscent of Illyrian tombs, of the Koman culture, but no traces were found to prove this hypothesis (**Figure 3a**).

3.2.3 Church at "Guri i Madh"

According to stories of residents, it is believed that even at "*Guri i Madh*", altitude 1212 m, there was a small Church [11]. Field survey confirms the existence of several oaks, tall pines, tracts of stone walls and a clandestine excavation under the oak, near the Great Stone. No ceramics or any other objects were found from the visual excavation. In the North of the area, pottery was found on a possible tumular hill. For this sub-area, information is scarce and should be looked at more carefully in the future.

(a) (b) (c)

Figure 3.
(a) View from the area with the supposed graves. (b) View of the Kuqesh walls, below the water sources, 2024. (c) Coral fossil of the Species Cunnolitеselliptica. *[Source: (a) Orthophophotos J.Pekmezi; (b) Foto E.Hoxha; (c) Found by J.Pekmezi].*

3.2.4 The water sources of Kuqesh

The area of Kuqesh, below the sources of drinking water, which supplies the village of Lubonjë, seems to be the richest area with archaeological finds, which are ceramics, construction material, fragments of walls and foundations. In this area, about 70 ceramics from the following periods have been found: Late Bronze, centuries thirteenth to twelfth BC; Iron, centuries eleventh to nineteenth BC; eighth to seventh centuries BC; 750–650 BC; Late Antiquity, fifth to sixth Centuries BC; Precivil 650–450 BC; and Modern period. The detailed analysis of the orthophotos, DTMs, DSMs and 3D Models obtained from the data processing provided by the Drone survey made it possible to detect several terrain anomalies, linear and circular. These anomalies led the team to suspect that there could be longitudinal/linear or circular objects hidden by vegetation in these areas. After accurately determining the location of the anomalies, the team returned to the field. From the on-site verification, two walls were discovered: one with a length of about 10 m and a height of about 1.5 m and the other with a length of about 80 m and a height of about 1.5 m. The first wall is thought to be the wall of a small settlement, and the long wall as a retaining wall for the terrain, or as some other platform. Ceramics were found along the entire length of the wall/levee wall (**Figure 3b**). The project founds also an interesting coral fossil of the Species *Cunnoliteselliptica (Cyclolites ellipticus)* of Upper Cretaceous age in the *Kuqeshi*. The condition of the fossil is somewhat unusual compared to common fossils, where the specimen is covered by a crust of iron oxide, with a reddish brown color (**Figure 3c**) [12].

3.3 Guri i Shqipes

Guri i Shqipes is located above Breza into the west of Kuqesh. It is a steep peak, at an altitude above 1300 m, almost bare, except for some tall pines. It has been a resting place for birds, especially eagles. This is also one of the reasons for the name. It is thought that this peak "...may have served as a Fortress or Fortress or Lookout point" [9]. This hypothesis is also supported by the fact that there are ruins. During our survey, it turned out that there were ruins on the top, as well as ceramics, fragments of dishes from the periods Bronx centuries fifteenth to twelfth BC; Iron Age, Centuries eleventh to twelfth and eighth to sixth BC; Late Antiquity centuries sixth to fifth AD. The analysis of the findings shows that *Guri i Shqipes* may be the earliest settlement referred to in the data so far, but that should be verified and compared with other areas in the future.

3.4 Breza

Breza is a very interesting area, considered one of the early settlements and one of the main neighborhoods of Lubonja. Situated at an altitude of about 1200 m, near *Kuqeshi* and above *Mëhalla e Sipërme*, Breza has a favorable position because it is where the roads that come from Lubonja, Kuqeshi and *Mëhalla e Sipërme* intersect with the road that continues to Castle, Psari i Zi, Mali i Zi, Selenica e Pishës and beyond. In terms of tourist attractions, Breza also has an artificial lake and a source of drinking water, which is a very suitable place for tourism. According to the legends, it is believed that a very fierce battle took place in this area, where many people were killed. It is not known which are the parties that have fought and in what period it happened [11]. From a quick visual survey of this area, ceramics and fragments of walls were found at the elevation of about 1195 m, while from the study of orthophotos and drone images, some hidden ruins were noticed among the forest, which should be looked at in the field in the future (**Figure 4a**).

3.5 Mëhalla e Sipërme

This area includes: the part below Breza; the artificial reservoir; the "*Saint Marine*" Church; the place called "*Varret e Evgjitëve*"; *Mitrogjoni* area and goes down to the area called "*Ledhet*" (**Figure 5**). Pottery has been found from previous surveys in this area, and the ruins of the old settlement were clearly identified during the project. The church does not exist today, but there are ruins and centuries-old oaks. Ceramics from the periods of Bronx, centuries fifteenth to eleventh BC and Iron, centuries eleventh to nineth BC were also found under the church.

3.6 Castle of Lubonja

3.6.1 Castle location

The Castle of Lubonja is located at the intersection of the border of the villages of Lubonja, Boshanj and Psari i Zi. Because of this position, it is known as Lubonja

(a) (b)

Figure 4.
(a) View from Breza, walls and ruins. (b) The position of Lubonja Castle in relation to the Kolonja plateau. [Source: (a) Drone photo, Jeton Pekmezi, 2024; (b) Photo: Neritan Ceka].

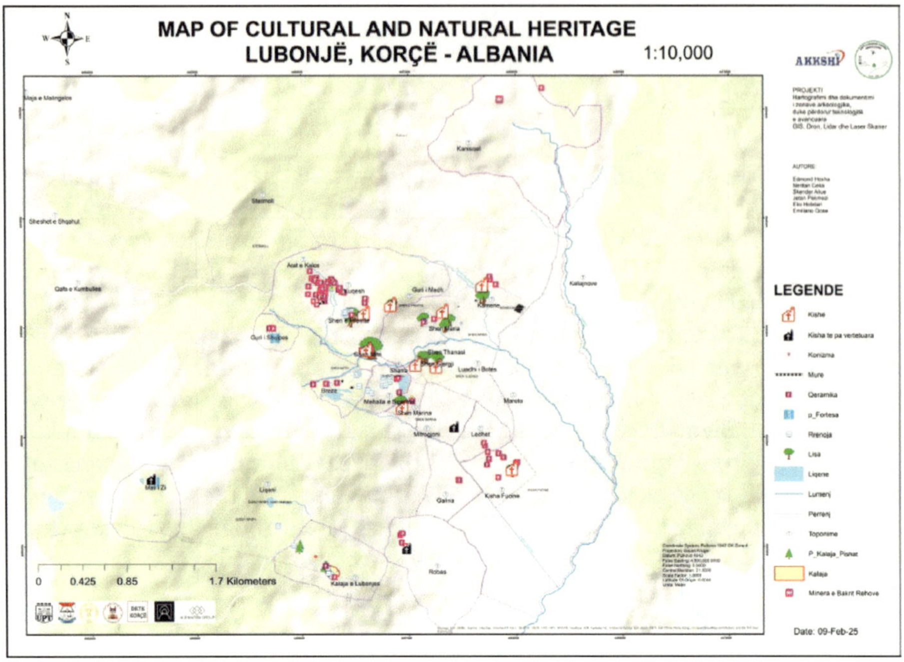

Figure 5.
GIS map of the cultural and natural heritage of Lubonjë village, Korçë 2024 [Source: HDZA Project].

Castle [13, 14], but in one case also as Boshanji Castle [15, 16]. Based on the existence of the tower, which was described as circular (**Figure 6a, b**) and the technique of building walls with large unhewn blocks, previous researchers have dated this castle to the centuries third to second BC [15]. In order to clarify the current situation, a new view of the fortification was carried out in the framework of the project. Initially, the re-evaluation of its position in relation to the treve (area) where it was

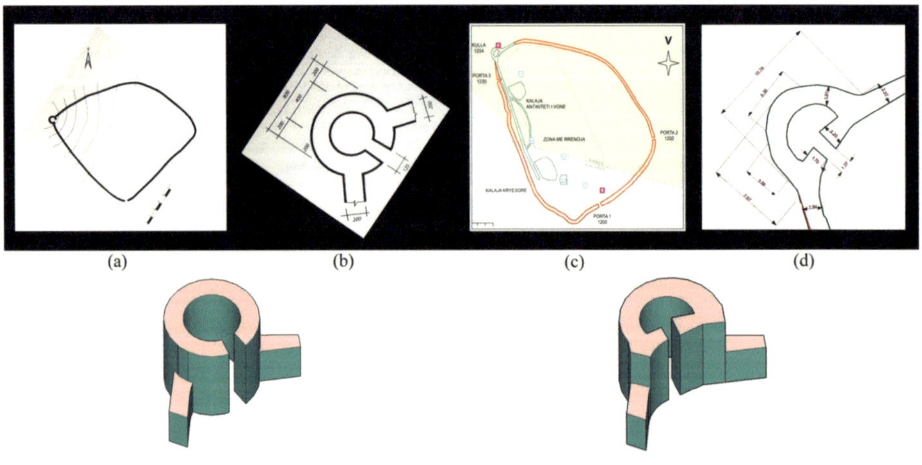

Figure 6.
The plan of the castle and the tower. (a) The plan of the castle 1981; (b) The plan of the Tower according to Karaiskaj 1981. (c) The plan of the castle from current data 2024; (d) The plan of Tower from current data 2024. [Source: (a, b) Karaiskaj Gj. [17] (c, d) N.Ceka, E.Hobdari, S.Aliu, E.Hoxha, J.Pekmezi].

located, as well as a more detailed study of its construction elements, was carried out. In order to clarify the period of construction of the castle, in the framework of the evaluation of the cultural heritage of the village of Lubonja, the HDZA Project made a re-evaluation of this archaeological site and in this study the Castle is named the *Castle of Lubonja*.

3.6.2 The positioning of fortification

The ancient fortification was erected on one of the heights of a ridge that descends from *Mali i Zi*, toward the valley of the *Osum*i River. The hill on which the fortification is erected is separated from that ridge by a higher peak on the western side, from where it scales toward the East with several terraces suitable for living. The area within the perimeter of the fortification is heavily wooded and has a pronounced slope of about 30° and the single tower is located at its highest point at the altitude of 1234.5 m at the top of the preserved walls.

The hill is equidistant from the agricultural lands that open at the bottom of the *Mali i Zi* slope between the villages of Lubonja and Boshanj, as well as from the pastures of the *Mali i Zi* slopes. More important is the fact that the hill ridge, where the fortification is located, offers a natural passage from the northern part of the Kolonja plateau through *Lubonja* to *Panarit* and *Qafa e Martës*, then through the valleys of *Osumi*, or *Devolli* toward the coast. The traces of the ancient fortifications are followed along their entire length, covering an area of about 1.2 ha, roughly taking the shape of a right-angled triangle with walls 405 m long (**Figure 7b**).

Surface observation was first carried out following the lines of the fortification walls and extending to both flanks, as much as the vegetation allowed. The walls, starting from the two flanks of the tower, descend toward the South-East following irregular lines, mainly dictated by the configuration of the terrain. On the western flank of the castle, it is possible to trace part of its length to two walls, one outer, lower down the slope of the hill and one higher up inside the perimeter. This inner wall is lost in about 2/3 of the length of the castle. The southern and eastern parts of the outer wall are better read and easily followed. In some places, it is well preserved in height, making it possible to read the construction technique.

(a) (b)

Figure 7.
(a) Peak with a height of 1234 m from the neck that separates it from the hill ridge. The pine trees of the Lubonja Castle tower. (b) Map of Lubonja Castle, 2024. [Source: (a) Photo: N. Ceka; (b) E.Hoxha, J.Pekmezi, N.Ceka, S.Aliu, E.Hobdari].

3.6.3 Periods of fortification

The fortification starts at the highest point with an elevation of 1234 m on the northwest side of the hill. From here, the side of the leg of the right triangle follows almost a straight line, while the northeastern side, of the hypotenuse, curves along the slope of the terrain, to join the final leg at the height of 1200 m. This shape is completely different from the quadrangular one of Karaiskaj's publication [15, 16]. The terrain included within the walls scales in the form of natural terraces up to a limited square, which rounds off the surface in its lower part. The town was easily reached in its upper part, where a low neck separated it from the neighboring height of the hill ridge (**Figure 7a**), but the south-western and northeastern sides protected by the walls constitute pronounced slopes, difficult to cross. Even the south-eastern side, at the end of the town, also has a significant slope that makes it difficult to climb toward the town. Despite the natural defenses that presented part of the terrain, the town has a complete walled enclosure. From the first observation, the two construction periods of its fortification are clearly distinguished in terms of construction techniques and protective elements, despite the fact that for the most part they have the same line of walls.

Starting near the upper gate (Gate 3) (**Figure 7b**), outside it, the outer wall of the fortification begins to appear, which is followed around the entire perimeter until it joins, in an indistinct way, with the tower. This is built with rough stones of the area laid in dry. To the north of the tower the beginning of the fortification wall is not clear. There, the wall is covered with earth; it looks like a levee/wall. It is impossible to distinguish an outer and an inner perimeter. The northern part of the fortification wall is preserved at a considerable height, even over 100 cm from the current quota of the breach. According to the testimonies of the residents, that part has been preserved even higher. This wall preserves no traces of binding material (lime mortar) and seems to belong to the earliest fortification. The state of preservation may suggest that this part (a tract of several tens of meters) may have been included in the new, smaller perimeter of Late Antiquity (see below for dating), but this remains to be proven.

3.6.4 The first period of construction (centuries eighth to seventh BC)

The earliest fortification is distinguished by its complete triangular shape, with long lines, with light arches, but without the use of right angles or towers. It begins with a wide arch at the northwest peak of the hill, from where the passage was controlled by the neck that connected it to the ridge of the hill. It then continues through arches to the leg of the end part, where the main entrance is also located. Throughout the length of the walls, the first period of construction is characterized by a similar construction technique, which is distinguished by the use of large blocks extracted from the limestone layers of the height to its northwest, where traces of quarries scattered on the hillside can be seen.

3.6.5 Walls

The blocks are placed on the wall without being worked on and the natural quadrangular shapes have allowed for a regular placement of them on the facade, taking care not to create gaps between the stones by filling the spaces between them with small stones, which gives the facade the appearance of a polygonal wall (**Figure 8a**). In the width of 2.30 m, the wall has been built on two sides, the outer and the inner, of large and medium blocks, as well as filling between them with small stones. The

(a) (b)

Figure 8.
The walls of the castle of Lubonja. (a) The structure of the double-faced surrounding wall and the fill; (b) The inner face of the surrounding wall on the eastern side of the fortification [Source: photo: N.Ceka].

wall is preserved up to a height of 1.80 m, especially on the northeastern side of the town, while the inner side is preserved up to the height of the terrace created inside it (**Figure 8b**). The little fallen material, in front and behind the wall, proves that it did not have a very high height. The facade, which testifies to the work of specialized people, has an inward slope to ensure its stability (**Figure 9a**). This technique is known mainly from the fortifications of the Early Iron Age, or the Proto-urban one [17]. The route of the walls is characterized by straight lines that fit the terrain with slight curves, never using right angles. Only in the three main corners of the layout does the track turn in wide arcs.

3.6.6 Architectural description of the constituent elements of the fortification

The castle was made by starting from the highest point, actually the access point to the fortification, and descending gradually to the lowest point. The castle of Lubonja (first period), according to the information so far has three entrances along the entire length of the route of the fortification walls, positioned approximately according to the three main geographical directions: South-East–West. *Gate 1*: From the information so far, this gate is the main entrance and is located in the lowest part of the fortification of the first period, near its southwest corner. It represents a

(a) (b)

Figure 9.
(a) Inward slope of the enclosure wall on the east side. (b) Gate 1 of the Castle. [Source: Photo: (a) N.Ceka, (b) E.Hoxha].

simple passage with a width of 240 cm where larger but palatable blocks were used and placed in more or less regular rows (**Figure 7b, 9b**). The entrance is simple, like a break-in-the-wall track. The shoulders consist of two rows of massive rough stones with a height of 50–55 cm. There is no trace of the placement of the door shoulders. Even in their continuation, the wall is mainly built with such massive stones. The masonry is dry, as are the other parts of this fortification. It was not possible to observe the foundations of these walls to have more complete data on the construction method. Two rows of blocks with a height of about 50 cm are stored on her shoulders. The clearance showed that the quota of terrain there was almost that of the former sole. This gate, with changes in topographic positioning, is also shown in the plan of the Castle presented by previous studies (**Figure 6c**) [15, 16]. *Gate 2*: The second entrance is found on the eastern side, which is also an interruption of the wall track but only 1 m wide. This gate was not part of this study (**Figure 7b**). *Gate 3*: The third entrance is located on the north-west side of the castle, about 15 m from the semi-elliptical tower, and constitutes a narrow passage of 0.80 m. Even in this case, the use of blocks with close quadrangular shapes is noticed (**Figure 7b**). The gate next to the tower clearly belongs to the second fortification. In its upper wing (the left side when viewed from the outside), the mortar of the wall can be seen. The shoulders are built with typically large stones, including those with a length of over 100 cm and a width of over 50 cm. Both shoulders have sloped inward, perhaps from the influence of vegetation, or from the weight of the walls themselves. Today, the walls are preserved about 90 cm above the ground level, but in the past, they must have been at a higher height. The opening of the gate in the field quota is 110 cm, while the upper part of the walls is only about 90 cm. It cannot be estimated how high this present elevation of the ground is, from the elevation of the original breach, nor can we imagine what the opening of the gate is at the base of the walls. Near this gate, to the east, below the Tower, a ruin is clearly evident, where the corner of a rectangular building is also visible. It seems that there was also a clandestine check there. Based on the above: the features of the plan; the type of masonry, with an original height of over 2 m; with a tendency to align the blocks on the facade and fill the gaps between them; the simple frontal type of the entrances; and the fact that we have no other data, either from random finds or from archaeological surveys, give support for the dating of the fortification of the first period of construction in the Second Iron period, in the centuries eighth to seventh BC.

3.6.7 The second construction period (fifth to sixth centuries AD)

In the northwestern, higher part of the fortification, a second construction period can be clearly distinguished, during which it was intended to use the earliest routes of the walls for the construction of a new fortification. Its strongest point was the construction of a Tower that controlled the low neck of the northwest side, which was placed on the arch of the walls of the first period of construction, completely covering it. From a second, more detailed observation made by N. Ceka and E. Hoxha inside the surface of the castle, it appears that it is possible that the late castle was built in the upper part, approximately two-third of the surface occupied by the earlier fortification. The main elements of this castle are: Tower, Walls, Gate 3, which, according to the data so far, seems to have been used by both fortifications and the ruins present on an area of about two-thirds of the main castle shown in the Plan of the Main Castle with a rectangle with broken lines (**Figure 7b**).

3.6.8 The tower of late antiquity Castle

The only fortification tower is in its uppermost part, where, by means of a neck, the fortified terrain is connected to the slope of the hill range (**Figure 10**). The tower of the Late Antiquity Castle is placed above the wall of the first period and has a semi-elliptical shape (**Figure 6c, d**), quite different from the one early published (**Figure 6a, b**). Perhaps at that time it may have been more crowded and perhaps the walls were more hidden. From the survey, it seems that there was a clandestine digging intervention in the tower. This has made it possible for the structure today (2024) to be more readable, and at the same time, two deeper pits, one in the northern corner and one at the entrance threshold, allow us to sufficiently read the deposits inside it as well as the construction technique in the lower part of the walls.

3.6.9 The stratigraphy

It is simple, about 70–75 cm of the lower part are remnants of mortar from the collapse of the walls and on it there is a thick layer of stones that came from the demolition of the masonry with a thickness of 130–180 cm with irregularities caused by the intervention made. *The Plan (layout)*: From a planimetric point of view, the tower is placed above the wall of the first period and has a semi-elliptical shape with a base of 7.6 m and a projection of 10.8 m. The tower is arched extending slightly more than half a circle (196°). This is because the wall that serves as the chord of the tower, and where the entrance to it is located, is not straight but broken right in the middle of the length. It has slight deviations from the regular circle segment shape. The inner diameter is about 5.6 m, while the outer one is about 9.7 m with a wall thickness of 2 m. The cord wall is about 2 m thick, while the access gate to the tower has an opening of 127 cm (**Figure 6d**).

3.6.10 Tower walls

The tower walls are preserved to a considerable height, with almost the entire height not exposed. The construction technique is irregular with unworked, or minimally

(a) (b)

Figure 10.
View of the Tower. (a) State of the tower before cleaning and (b) The condition of the Tower after the end of the expedition. [Source: Photo: N.Ceka, E.Hoxha].

(a)

(b)

Figure 11.
(a) The holes of the floor scaffolding beam on the west wall of the tower. (b) Work moment in the tower hole, 2024. [Source: Photo: N.Ceka, E.Hoxha].

worked, stones not placed in regular rows and bound with quality and relatively abundant lime mortar. There does not seem to be any tendency to place large stones in the lower rows and small ones above, at least for the inner face of the wall. The stones range in size from a maximum of 75 cm x 30 cm to as small as 10 cm x 10 cm. In the inner arched side wall, two holes for scaffolding beams for the placement of a floor can be seen. This fact proves that the tower was two-storied, with a kind of warehouse on the lower floor. The holes are located opposite each other, about 60–70 cm away from the corners where this wall meets the straight parts of the tower chord. They are not at the same level but have a difference of about 30 cm. The depth could not be measured exactly (**Figure 11a**).

The holes made in the northern corner of the tower and in its gate were slightly cleaned. They were filled with some fallen soil. At both points, it seemed to have reached the floor, in the first and on the threshold of the transition in the second. In this way, in the northern corner of the tower, the floor was covered with the same mortar as that of the wall, laid on a layer of small stones. It was stored a little at the foot of the cord wall. It seems that the arched wall of the tower had a plinth that partly protruded quite shallowly or merged with the wall. It was not excavated deeper, and neither was the Project intervened. Some ceramic fragments were encountered here (**Figure 11b**). The finding of this *in situ* vessel of the pottery type on the ground floor confirmed the idea that the ground floor could be a storehouse. Even the lower part of the wall at a height of about 50 cm, had the joints well leveled with abundant mortar, and in some parts, even the stones were completely covered by this mortar (**Figure 11**). The tower and the (inner) fortification wall in its southern part, where the third identified gate is (**Figures 6c, 12a**), were built with unworked and some possibly slightly worked stones bound with good quality and relatively abundant lime mortar. Below the gate, this wall is identified as a line, but its technique is not discernible. It should be considered that the technique continues the same below. The walls of the tower are 2 m wide and are built of small to medium-sized stones bound with hard lime mortar with coarse stream sand binder. It has an entrance with a width of 1.27 m, but with two wings unequal in width and length, respectively 2.20 and 1.70 m (**Figure 12a**). The tower was not excavated and not all cleaned because the rights to this project were limited. Likewise, the walls were not revealed to the depth of the sole both internally and externally. This remains the task of ongoing projects.

3.6.11 The walls of the late antiquity castle

Two walls of equal width of 2 m separate from the tower, passing over the tracks of the first construction period. They are built in the same way as the tower wall, being

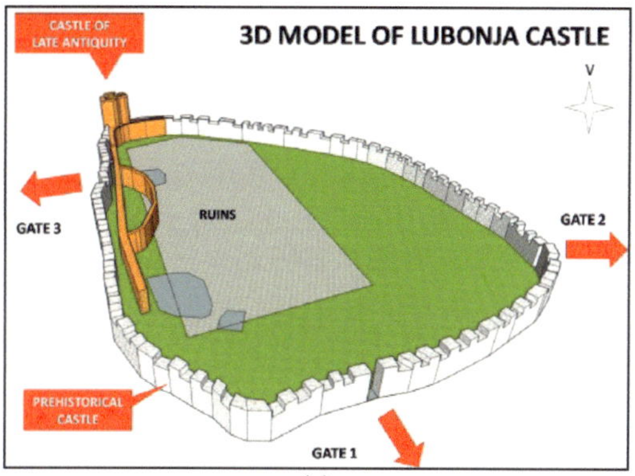

(a)

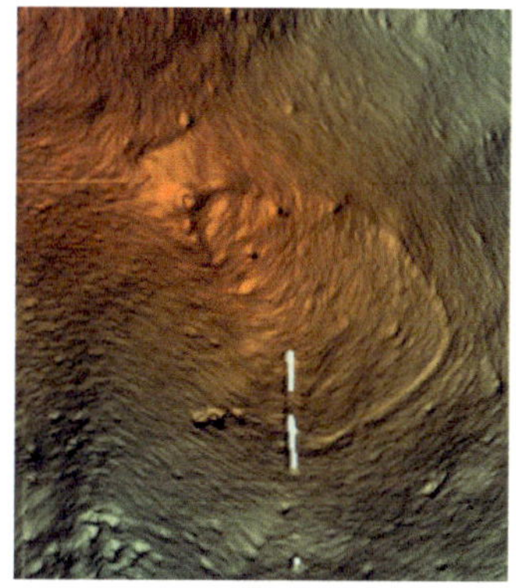

(b)

Figure 12.
(a) 3D model of Lubonja Castle, 2024; (b) View of the walls of the Castle, Lidar. [Source: E.Hoxha, J.Pekmezi, E.Qose].

placed directly on top of the first-period wall but using a weak mortar that has been crushed between the stones. The wall of the western side is more clearly followed after 14 m, where in the same direction as Gate 3 of the first period, an entrance with a width of 1 m is quite well preserved, with a simple construction where it is not clear that there was any gate rabbet. From there, the track of the second period is followed in the form of a straight line until about 77 m near the track of the end wall of the first period (**Figure 7b**).

3.6.12 Gates of late antiquity castle

From the data so far, it appears that the Late Antiquity Castle has one Entrance. It is located in the same direction as Gate No. 3 of the Castle of the first early period. This entrance is quite well preserved; it is 1 m wide, with a simple construction (**Figure 13a**) where it is not clear that there were any gate rabbet. From there, the track of the second period is followed in the form of a straight line until about 77 m near the track of the end wall of the first period. Its proximity to the tower and its location on the edge of the hillside prove that it was not the main entrance. From

(a)

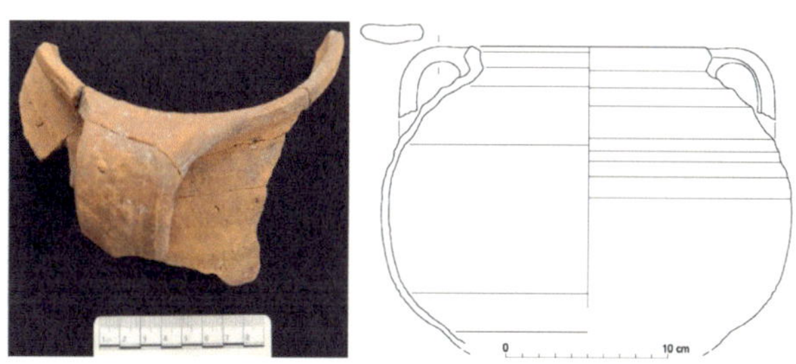

(b)

Figure 13.
(a) The Entrance of the second construction period. Castle of Late Antiquity. View after cleaning, June 2024 (b) The jerry (vorba) partially rebuilt Jerry. [Source: (a) Photo: E.Hoxha; (b) Elio Hobdari].

there, the route of the second period is followed in the form of a straight line until about 30 m near the route of the end wall of the first period (**Figure 7b**). In this study, due to the forest and dense vegetation, it was not possible to follow the continuation of the eastern wing of the fortification of the second period, to pinpoint the place of return toward the western wing. One possibility is that the builders of the second period reinforced with new walls only part of the fortification, adding the tower and its wings and leaving the well-preserved walls of the first period in use.

3.6.13 Buildings within the castle of late antiquity

A group of buildings with walls of small stones without mortar are associated with the second building period near the western wall. One of them, an arch with a length of approximately 40 m and a radius of about 10 m, was supported by the surrounding wall, forming a kind of shelter. In the Plan of the Castle (**Figure 7b**), this area is shown arched in green marked with (possible accommodation). Likewise, in front of Gate No.3, a more or less square building can be seen. The corner of the wall can be seen on the ground ruins.

3.6.14 Archaeological material

Two types of archaeological materials have been found in Lubonja Castle from the data so far: (1) ceramic vessels and (2) stones used for grinding grain. The ceramics were found in the Tower, and the grinding stones near Gates 1 and 2. The fragments of ceramics found in the open pit inside the tower are the only ones that were obtained from the entire surface survey. All the fragments belonged to a jerry (vorbë) dated to the sixth century. The jerry (vorba) is one of the most common kitchen utensil excavated at Late Antiquity sites. The vessel was partially rebuilt. Of the two urns found, only one was joined, but the other probably also belongs to the same vessel (D: 15 cm; h: 18.5 cm (preserved). Rim slightly protruding. The blade starts at the top of the rim. Section of the blade: 2.8 cm x 1 cm. The bottom is not preserved, but must have been rounded. Wall thickness 0.5–0.7 cm. Circle grooves prominent in the shoulder part. Stones used for grinding and building materials were found during previous surveys by E. Hoxha and S. Aliu in 2021 near, above Gate No. 2.

3.6.15 3D model of Lubonja castle

The 3D model of the Castle of Lubonja was created using data obtained from drone flights, GPS Surveying, Total Station and LIDAR data. The programs with which the model was built were Pix4D for the data obtained with the Drone and Modeling and realization in the program Sketch UP and Global Mapper (**Figure 12**).

3.6.16 Chronology and dating of the fortification

Second Iron Age Fortification (Initial Castle) centuries eighth to seventh BC). It stands out for its complete triangular shape, with long lines, with light arches, but without the use of right angles or towers (**Figure 7b**). The entire length of the walls is characterized by a similar construction technique, which is distinguished by the use of large blocks extracted from the limestone layers of the height to its northwest, where traces of quarries scattered on the hillside can be seen. Referring to previous researchers have described this Castle as possibly prehistoric, giving the possibility that it belongs to the centuries third to second BC; they even doubted the late antiquity [15, 16]. Based on the complete observation, in the framework of this Project (2024), of the early Castle, the cleaning and clearer discovery of its walls, the clarification of the overlap of the walls of the second Castle on the early fortification, and moreover the shape of the Tower and its placement, which is completely different from what the previous authors describe, the authors of this study not only eliminate the doubt of the previous authors, about the Late Antiquity as the period of its construction, but place the Initial early Castle in the period of Second Iron Age centuries eighth to seventh BC.

In both periods of its life, based on actual data, the Castle of Lubonja represented a fortification with the same function, which was the control of the natural road that passed from the plateau of Kolonja to the valley of Osum and further to the Adriatic coastal plain. It seems that in the early Iron Age, in c. XI-VIII BC, that route was important for livestock based on transhumance, when the summer pastures of the Kolonja plateau and the Gramozi Mountain may have been the migration site of the flocks that wintered in the coastal plains. Securing this road during the journey may have been the main function of the Castle of Lubonja, since the surface observations do not seem to have had an intense life as well as time continuity for the ancient period. However, based on the data so far, it seems that the Castle of Lubonja did not manage to develop into an urban center in the Illyrian Civil period, when the capital of ancient Cologne, the territory of the Kaloikins, was in Hollmi e Qinami. However, it remains to be tested in the future.

3.6.17 Late fortification (late antiquity castle)

For the period of Late Antiquity, in the historical context, Lubonja Castle represents a system of fortification created during the c. IV-VI AD in the conditions of the crisis of the Roman Empire and the passage of this territory under the Byzantine administration. The significant number of fortifications of that period on the plateau of Kolonja and of Korça is related to the role these areas took for the security of the roads that connected Constantinople and Thessaloniki, as the main center of the Byzantine administration with the Adriatic coast and Dyrrah as the main center of Byzantine Illyricum. Even in this system, Lubonja Castle had a role in controlling the natural road that connected the Kolonja plateau with the middle valleys of Osum and Devolli, where the ancient city of Partha (Berati) played an important role. On the other hand, the construction techniques evidenced in the tower and in the upper part of the wall next to it, as well as the ceramic material, sufficiently clarify the dating of the second fortification in late antiquity.

The well-curated construction, given the available building material and especially the abundant lime mortar of good quality, place this construction in a time segment that does not go later than the middle of the century sixth AD. Therefore, a date between the century fifth AD and the first half of the century sixth AD is the one that can be proposed at the current stage of knowledge. In the absence of archaeological excavations that would eliminate the ambiguity, we can assume that the wall in the upper part of the Late Antiquity fortification, in the part of the tower, was built on top of the early wall, while the tower stood outside this perimeter. This would explain why the chord wall of the tower is not straight but broken at an angle in the middle of it (right where the entrance gate to the tower is). The fact that we have no cracking of the arched wall of the tower at its extremes, as usually happens in such construction schemes, can be explained by the strong foundations of the new construction (to be verified), or by the fact that it was built on solid rock ground (also to be verified, but more possible). Below the tower, to the south, the wall took an immediate turn, entering the old perimeter (where gate No. 3 is next to the tower). In the northern wing, where it was not possible to observe anything more than the brick that was visible on the surface, it is very possible, also seeing the direction of the chord wall of the tower, that the wall of late antiquity was superimposed on the early one. There are many possibilities that in the northern part both lines of the wall are superimposed. In that partly rocky terrain and characterized by large masses of wall stones lying on the surface, we think the only way to fully investigate the fortification would be with surveys precisely distributed along the perimeter.

In the general assessment of Lubonja's cultural heritage, the Castle is the main asset for the period of Illyrian and Illyrian-Byzantine Antiquity. In the context of the prehistoric fortifications of the Early Iron Age, it presents a special interest in determining the chronology of the early Illyrian fortifications. It is at the same time a tourist asset for the exceptionally good state of preservation of the fortification and with a little cleaning and careful restoration it can become a center for tourists to visit in an area that would include the Church of the "*Dormition of Saint Mary*" established on 1015, and other monuments as well as the Natural Monuments identified by the Project.

3.7 Rrobas

Referring to the geographical position, the settlement of Rrobasi is located on a steep terrain at the end of Mali i Zi. The toponym Rrobas is an early source name that belongs to the lexicon of today's Albanian language. The Rrobas part of Lubonja, with fertile lands for agriculture, with hills covered with wood and pastures for livestock, protected from the cold winds of the North, has been inhabited since early times, according to the statements of the people of Lubonja. In ancient history, based on the stories of the villagers and some written sources, it is thought that Rrobas was a very early settlement. According to legends and M. Xhaxho, Rrobasi was very big, with 6 neighborhoods (*mëhalla*) such as *Rrobasi*, *Galina*, *Ledhet*, *Mitrogjoni*, *Shën Marina* and *Mëhalla e Sipërme* [18]. During agricultural works, copper kitchen utensils were found, which are thought to be very old. During the visual inspection of the project, ceramics and a tract were found, which may be the foundation of a wall. However, these data are not sufficient to reach a clearer conclusion for this area, so it is necessary to undertake control surveys. *Galina* was considered as neighborhood of Rrobasi and is separated from *Ledhe* by the "*Galina stream*". It is not known yet why he got this name. This area is notable for its many ceramic finds, especially bricks and tiles. The discoveries, by the villagers, of some fireplaces have led to the opinion that there may have been a workshop for the production of bricks. During our expedition, ceramics and pieces of bricks were found in the fields and especially in the ruins, which support the existence of such an activity. *Ledhet (Levee)*; they are thought to have also been a part, or neighborhood, of Rrobasi. It seems that this area got the name "*Ledhe*", because of the relief and the soft soil at risk of erosion. For this reason, this area has many retaining walls, or walls (**Figure 5**).

3.8 The field of Lubonja, the church "*Kisha Fuçinë*"

The field of Lubonja is another prehistoric settlement of Lubonja. During surveys in the field and along the drainage channels, the project team found and collected several ceramic fragments of different dates from the second period of the Late Bronze c. XIII-XII BC and continuously during the Iron Age c. XI-VII BC, Roman period c. I-IV AD as well as many fragments of Late Antiquity c. V-VI AD. What is worth noting is the existence of piles of stones in the form of mounds, but this fact must be proven in the future. Among them there are also several mounds that have been elevated with earth, through burials, one of which was partially damaged by treasure hunters years ago. The Church "*Kisha Fuçinë*" is thought to be a tumulus cemetery located in a field. Local residents perceive this cult object as an important holy place in the old days. It is thought that there was also an old cemetery there. All the time various works have been done in the field, but this place has never been touched or damaged. During the survey from the very first moment in the South-western part at the entrance of the Church, near the great Oak, a pit opened by clandestine treasure-seekers was found. At the bottom of this pit, a wall tract with two lines of stones was found, measuring 120c m x 40 cm, a fact that makes us think that it is possible to deal with a church of graves. During these works, some pottery fragments dated to the Late Antiquity, c. V-VI AD.

3.9 Mali i Zi *(Black Mountain)*

The highest peak of Mali i Zi is located at 1703 m. Perhaps the name originates from the fact that the mountain is filled with dense fir forests that look dark from a distance. According to M. Xhaxho "*... there was a church in Mali i Zi*" [19]. The survey focused on two areas. Visual inspection revealed random pottery and an arched wall/ foundation in the first area. Unlike the first area, which has no high vegetation, but only grass and various flowers, the second area has more trees and mainly oaks, which are not very tall. In this area, the presence of many clandestine excavations is noticed. This area has the most promising data for finding signs of habitation. No ceramics were found on the surface in this area. From the conversation with the villagers of Lubonja, there is no more approximate data on where the Church could have been (**Figure 5**).

3.10 The mountain of Stërmolli and Kalidhe

A quick surface survey was carried out in Malin e Stërmolli. Villager's stories speak of the existence of old settlements, but from this survey, apart from some vague ruins, no other evidence could be found. However, it is an interesting area to review in the future. Mount Stërmolli has an altitude of 1500 m, it is located above Kuqesh, and there was definitely a connection between them in the early past (**Figure 5**).

3.11 Industrial mining area

The industrial area of the Rehova copper mine, a very old mine, was not part of this study, but the stories of the residents and mine workers speak of the existence of the smelter, where traces of slag were found. Ceramics, complete vessels, were also found near the mine area. The fact that the Rehova mine is the only copper-bearing area in the whole of South-Eastern Albania makes its study very interesting.

Based on archaeological findings, historical materials, legends and oral stories, the HDZA Project modeled a complete rough picture of what Lubonja might have been like in its early history. The illustration clearly shows all areas with settlements and archaeological potential (**Figure 14**).

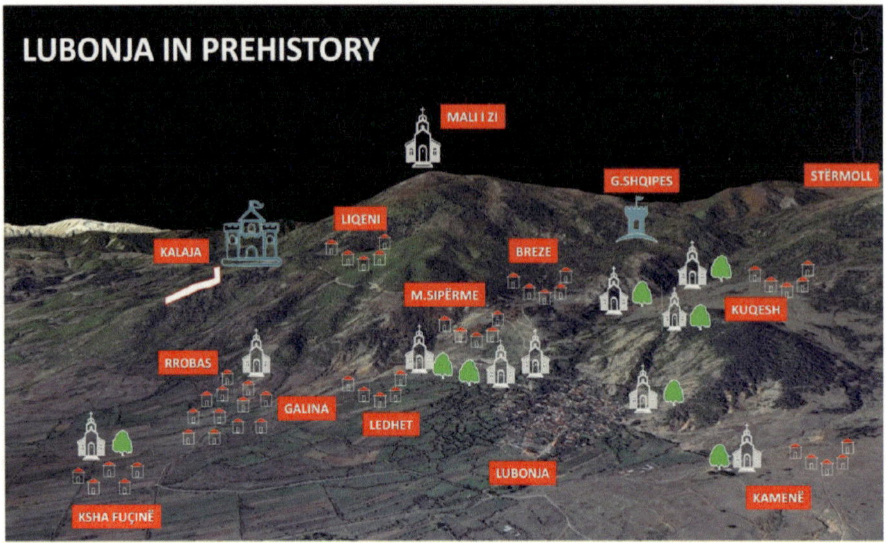

Figure 14.
Lubonja village in early history. [Source: Google Earth image. Modeling and illustrations by Edmond Hoxha].

4. Mapping and 3D modeling of cult objects and wall paintings using laser scanner

4.1 Modeling the church "Dormition of the Virgin" in Lubonjë, Korçë

GPS, Drone and Laser Scanner technology were used during this study in order to complete the survey and build 3D models. The object of the study was the "*Dormition of the Virgin*" Church. The church was built in the northwestern part of the village, located above the main road, on a slightly sloping ground, once surrounded by high walls. This church was built in 1015 [20]. The church was destroyed in 1806 and that same year it was rebuilt and painted again by the painter Tërpo Zografi. The second danger of this church was in 2004. This is the moment when the wall painting of this Church suffered the greatest damage. From a wall painting spread over all its pages, today we have only 1/4 of it, that is, only one painted page, which is also damaged.

4.1.1 The architecture of the church

The church is one-story, basilica, with arcades (columns), it is three-nave (with three internal divisions of the naos, the main part) with narthex and exonarthex, which in today's state is quite transformed. The church has a ceiling and a size of 26.8 m x 9.96 m x 6 m, a building area 286m^2, perimeter of 78 m and the total area of the area where it is located is about 1000m^2. In the narthex (part of the corridor) there was a separate room that communicated with the nave through a window. This room was for women and those who were not baptized, but who followed the liturgy (mass) from this window. From the foundation until now, as a result of various restorations, the architecture has undergone changes such as: The halls have been simplified and the original structure has changed. Thus, after the restoration in 2012, we have one arch less than the original. The original height of the Church has changed which has brought about an essential esthetic change. The part of the Church where the altar is located, which is thought to have been the original part of the foundation year 1015, was higher than the second part which is thought to have been expanded in 1806 (**Figure 15a**). The shape of the original roof has changed. Likewise, the shape of the roof (South–North section) has also changed, from trapezoidal to triangular. A rough modeled view of the interior of the Church is given in **Figure 16**.

4.2 Mural painting of the church "Dormition of the Virgin" Lubonjë, Korçë

The church dedicated to the Dormition of the Virgin has been inherited nowadays with many structural and architectural transformations. It is a three – aisled basilica, with a three-faced apse, a western narthex and a southern portico, covered by a gable roof. The interior of the church was entirely decorated with mural paintings, which covered an area of ca. 180 m^2. Today the major part of the painted decoration is lost. It covers an area ca. 40 m^2 and is preserved in a very poor condition. The murals belong to at least two different painted periods and are represented mainly by compositions of the eastern wall and some scattered badly damaged fragments of saintly images in the southern and northern walls of the church. The first layer which remnants are identified in the eastern wall of the church has been the subject of a number of scholarly studies suggesting an approximate chronology from eleveth to thirteenth century (**Figure 17a**) [21]. This layer consists of fine-quality mural paintings rendered by well-educated masters and is linked with the byzantine phase of the monument. Of particular interest in the fresco decoration of the Byzantine

(a)

(b)

Figure 15.
The view of the Church "Dormition of the Virgin" of Lubonja, Korçë. (a) View of the Church in 1967, (b) View of the Church in 2024 [Source: (a) Photo State Archive, (b) Photo Edmond Hoxha].

Figure 16.
The 3D model of the "Dormition of the Virgin" Church in Lubonja, Korçë, built based on the information of the residents and surveying with a Laser Scanner. [Source: Sketch Up, Edmond Hoxha].

(a) (b)

Figure 17.
View from the mural painting of the "Dormition of the Virgin "Church. (a) First phase c. XI-XIII AD "Saint Mary of Evangelization"; (b) Second and Third stages of mural painting. [Source: Photo: E. Hoxha 2022, F. Prifti, 2024].

period is the examination of the placement of the image of Pope Sylvester among the bishops in the apse of the altar composition Melismos, which constitutes an unusual feature for this period. In the churches of the territory, a series of images of Saint Sylvester are preserved in the altar space but never in the apse itself. This special emphasis on Saint Sylvester, Pope of Rome, placing him among the first six or even four fathers of the Orthodox Church, must be explained also by the historical circumstances of the time when these frescoes were created [22]. However, the question of the dating of the murals of the Byzantine period and the clarification of painting layers in the apse also remains open.

The major part of the preserved painted decoration belongs to the post byzantine phase of the monument and, according to the donor's inscription, nowadays lost, which records the names of the individuals that contributed to the erection and the decoration of the church, the murals are rendered in 1806 by the hand of Terpo Zografi, member of the renowned atelier of Zografi and son of Constantine. The best-preserved compositions from this phase are the image of Christ as Man of Sorrows in the prothesis, Meso-Pentecost (Mid-Pentecost), Jesus and the Samaritan Woman at the Well, Sunday of the blind man in the upper register of the eastern wall, medallions of martyrs in the middle and narrow register, full length representations and the podea in the lower register of the wall (**Figure 17b**). The main focus of HDZA Project 2024 and the team involved with the documentation and study of the architecture and painted decoration of the church of Dormition of the Holy Virgin in Lubonja village is the nineteenth-century painted layer, through the implementation of modern technologies and methods, including visual inspection of the painting and church and mural decoration Laser Scanning.

On the other side, another important feature of the interior of the church was the wooden iconostasis, which icons are preserved and conserved in the storage environs of the Albanian Museum of Medieval Art in Korça. The restorers and experts of this institution enabled the identification of icons, which were not exhibited since 1967. The implementation of laser scanner in the documentation of the iconography of the monument, provided data and detailed images of the wall painting in its original dimensions and generated accurate data related to the wooden iconostasis and its remaining parts, on the basis of which it became possible its hypothetical reconstruction of its structure (**Figure 18b**).

(a) (b)

Figure 18.
Use of Laser Scanner. (a) Church Scanning Process with Laser Scanner; (b) View of the Iconostasis of the Church "Dormition of Virgin" Lubonjë, Korçë modeled based on the data of the Laser Scanner and the Icons located in MKAM Korçë, Albania [Source: (a) Alb Matrix Group Ltd., E. Qose. (b) Icons MKAM. Identified: F. Prifti, G. Proko; 3D model, E.Hoxha].

5. Mapping of the natural monuments of Lubonja village

5.1 The natural heritage of Lubonja

The nature of the Lubonja area gives the visitor a nice surprise and amazes him with the variety of landscapes, which change rapidly from the pit plain to the heights of *Mali i Zi* Mountain. To get to know this "museum" that looks like a large natural "amphitheater" because of the Lubonja Pit, the visitor does not have to travel long, just about 3–4 km. Despite the marked natural diversity, in the area of Lubonja, until now, no area has been studied and evaluated as a natural heritage. This is the first study and special care has been taken to research the connection of the natural monument with legends, religious and pagan rites and historical events with the aim of knowing and preserving our rich spiritual history and world, the rooting of our national identity and traditions, making the Natural Monuments more interesting for visitors.

5.2 Natural conditions and their role in the natural heritage of Lubonja

Lubonja is located in an area where several units with special morphological and morphogenetic features can be distinguished, which are Vithkuqi Highland; Pit of Kolonja [23, 24]; Pit of Lubonja (1) In the *Vithkuqi highlands*, the elevations above 1200 m have been shaped by the glaciations of the last phase, Würm (about 10,000 to 25,000 years ago) modeling numerous glacial forms such as cirques, troughs, spats and moraines. The magmatic composition of *Mali i Zi* Mountain conditioned the preservation of these glacial forms and lakes until our days. Of them, only two are preserved, which are moving toward the stage of swamping and complete swamping (**Figure 19a**). The drying up of other glacial lakes is related to the regressive erosion of runoff. (2) *The Kolonja Pit*; lies to the south of the threshold of Qarri, to the east and to the north of the highlands of Kolonja, to the west of Mount Gramozi, at altitudes of 800–1200 m. It is 20 km long and 4–9 km wide and is formed in a graben

(a) (b)

Figure 19.
(a) View from Lando's Square, once a glacial lake, now dried up and turned into a meadow. (b) View from Lubonja Pit. In the background the pit of Kolonja [Source: Photo P. Qiriazi, 2024].

structure. The evolution of the structure of this pit begins in the Pliocene, when the Kolonja pit was covered by an intermountain lake. (3) *Lubonja pit*; which is separated from the Kolonja pit through an ultrabasic structural threshold, with a relative height of about 100 m, has an extended NE–SW shape of about 7 km and a width of up to 1 km. On the ridge of the structural threshold, with a relative height of about 100 m, that separates it from the Kolonja pit, in places where molasses deposits are preserved, the same as those that fill both pits. This shows that once these pits were covered by the same lake (**Figure 19b**).

The climate of the Lubonja area is Mediterranean sub-mountainous and Mediterranean-mountainous. The hydrography consists of the dense network of tributaries of the Upper Osum, represented by small streams of clean and cold water, which feed into the numerous springs in the heights of *Mali i Zi* [25]. The soils are mostly brown. At altitudes above 1100–1200 m, there are forest soils. In general, the soils are quite fertile. The natural vegetation consists mainly of oak bushes, which represent the understory of the dense oak forests of the past. Continuous abusive cutting of tall trees, mainly oaks of the *Qurcus cerris and Quercus pubescens* species, degraded the oak forest in the form of sparse shrubs. In Mali i Zi mountain, there are conifer pine forests, especially black pine (*Pinus nigra*) and spruce (*Abies alba*), and broad-leaved associations, such as mountain maple (*Acer pseudoplatanus L.*), white dogwood (*Carpinus betulus L*), black dogwood (*Carpinus orientalis Mill*), vidhi (*Ulmus foliacea, Gilib*), etc. The animal world is diverse: reptiles, birds, mammals, such as: rabbit, fox, wolf, brown bear, wild boar, etc.

5.3 Methods of surveying, measuring parameters and studying natural monuments

Initially, based on existing studies, data on the peculiarities of natural conditions, early population and its evolution in the area of Lubonja were examined and documented. Further, cultural, religious heritage and economic activity in different periods were studied. Bibliographic information was verified in the field, and natural conditions were studied such as geographical position in relation to natural and social indicators; communication links over the centuries and currently with other areas; construction and geological structure; relief morphology and its morphogenetic form; water bodies and their features; characteristics of landscapes and ecosystems, plant types, plant floors, structure, plant associations and their phenological state and the causes of damage. Conversations were held with residents and experts of the area and through advanced methods and equipment, detailed measurements of the morphological parameters of the centuries-old oaks, the pines of the Lubonja castle and the glacial lake were carried out. The natural objects were surveyed and based on these data the Natural Heritage GIS Map was constructed.

5.4 The study of century-old oaks

The project studied 39 old Oaks, which are located nearby the churches of Lubonja. The area of these oaks covers an area of about 3.1 ha. The oaks are mainly exposed in the east. The slope ranges from 2° to 40°. The oaks are located at elevations from 920 m to 1186 m above sea level. The circumference of the oaks ranges from 1.2 to 4.6 m, while the height ranges from 4 m to 23 m. Crown condition is good. Oak trees have from 2 to 26 main branches. Data collected in the field were arranged in separate tables for each area. **Table 1** and **Figure 20** present an example of the documentation of the parameters and the construction of the GIS Map as well as the view of the oaks. *Values of Oaks*: Oaks are remnants of dense oak forests, which have existed and have numerous values: (1) *Scientific* that, mean biological, ecological, geographical, historical values, etc., of great value to biologists, ecologists, geographers, researchers of landscape evolution, ecosystems, archaeologists, historians, etc. (2) *Religious and spiritual* values related to the fact that they are located near churches, have been considered sacred, and have become part of the spiritual world of the inhabitants, being transmitted from generation to generation until our days. The oaks were considered inviolable, and many stories and legends were woven about them, which have great value in recognizing and preserving the great cultural and spiritual wealth and identity of the inhabitants. (3) *Touristic and didactic* values, which increase their touristic attraction and turn them into didactic natural research-scientific laboratories [25].

5.5 The centuries-old pines near the Lubonja Castle tower and in Kalidhe, Stërmoll

Near the tower of the castle of Lubonja (**Figure 7a**) there are six centuries-old pines, with large sizes where the diameter of the trunk goes up to 0.8 m, the diameter of the crown up to 9.7 m, while the height goes up to 13 m. Crown condition is good. Pines have from 15 to 25 main branches. The area covers an area of about 950 m2, has an S-L exposure, the average inclination of the slope is about 35°, and it is located 1230 m above sea level in the S-W of the village. They are mainly of the Pinus nigra type and make up the upper floor of the conifer forest, while around them there are rare bushes of this sub-forest.

The presence of young pines of small size indicates that the large pines are remnants of the former large forest.

5.5.1 The values of the centuries-old pines

The pines are remnants of the dense conifer forests that existed in the area of Lubonja and have numerous values: (1) *Scientific*: biological, ecological, geographical, historical, etc., related to the large size and early, centuries-old age. This makes them of great value to biologists, ecologists, geographers, researchers of landscape evolution, ecosystems, archaeologists, historians, etc. For the exact determination of this age, in the absence of specialized measurements (object of the second phase), it will be judged from the time of the construction of the Castle, which dates back

Group of Oaks	**Areal surface (Ha)**	**Altitude (m)**	**Geographical coordinates**	**Orthogonal coordinates (Centroid)**	**Areal exposure N,E,S,W**	**Slope inclination**	**Position in report with village**		**Distance from neighbor oaks group (m)**
5 Oaks	0.7	1005	Lat = 40°28′15.31”N Long = 20°36′47.98″E	X = 4,467,350 Y = 4,482,033	V-J	Aproximate 40°	In the end of North part of village		60 m
Lonely Oak	Areal surface (m^2)	Height land-top of the crown (m)	Trunk diameter (m)	Trunk Perimeter (m)	Crown Perimeter (m)	Crown Diametric (m)	Crown status	Number of main branches	Distance between neighbors oaks (m)
Oak 1	383	16	1.46	4.60	69	22.0	E mire	10	L1-L2 = 11

Table 1.
The parameters of the oaks in the Church “Dormition of the Virgin” in Lubonja, Korçë.

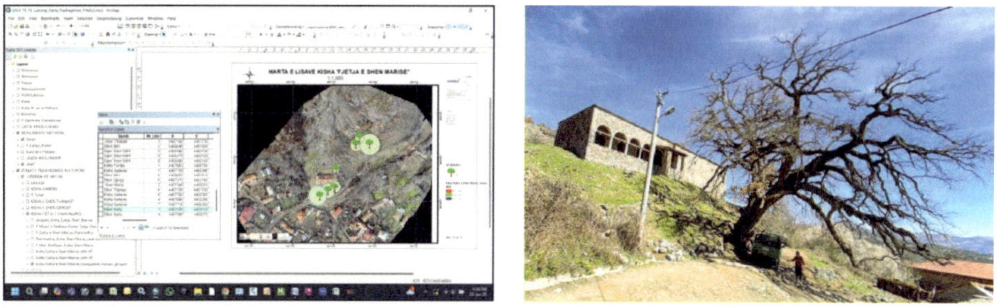

Figure 20.
GIS map of Oaks in the Church "Dormition of Virgin" Lubonjë, Korçë, 2024 [Source: GIS and photo: Edmond Hoxha, Orthophoto: Jeton Pekmezi].

to the centuries eleventh to eighth BC and centuries fourth to sixth AD. (2) *Tourist and Didactic Value*: related to their large size, centuries-old age, esthetic values, their preservation near the castle, which further increase their tourist attraction and turn them into natural didactic laboratories.

5.6 The lower glacial lake of Mali i Zi mountain

In this phase of the project, the Lower glacial Lake (*Gjoli i Madh*) of Mali i Zi was studied and mapped. In the **Figure 21** presented the map of the lake, which is located at an altitude of about 1300 m, and the central part of the water can be clearly seen on the map, with a circular shape with a diameter of 68.5 m, a perimeter of 215 m and a water surface of 3003m^2. The map also shows its marshy and swampy surroundings, with hygrophilous vegetation (reeds, etc.) with a perimeter of 273 m and an area of 4227m^2. The presence of lacustrine deposits and hygrophilous vegetation in the surrounding belt of the lake, which has a width of several tens of meters, speaks of

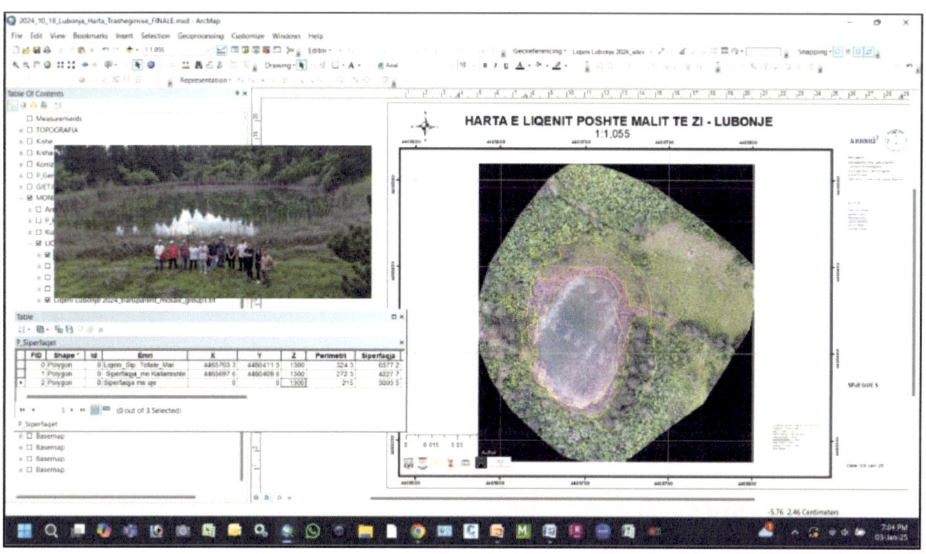

Figure 21.
GIS Map of the glacial lower lake of Mali i Zi, Lubonjë, Korçë. [Source: GIS: Edmond Hoxha, Photo and Orthophoto: Jeton Pekmezi June 2024].

the former much larger extent and the continuous reduction of the surface and its drying. This evolution toward the drying up of this lake is related to general, local, natural and anthropogenic causes. In our specific case, all local natural and anthropogenic factors have accelerated the drying of other lakes in Mali i Zi and risk drying up its last two lakes. Lower Lake, the case of this study, was strongly affected by the artificial opening of an illegal drainage channel about two decades ago, turning it into a surface-flow lake.

The unique values of the Lower Lake of Mali i Zi are expressed in The great purity of the water, the unique and diverse living world of plants and animals, perhaps with endemic creatures, *Great scientific importance* (geographical, hydrological, biological, ecological, etc.); *Didactic and tourist importance*; Stunning esthetic appearance; The lake is only a creature of nature and its laws; It is a remnant of the last glacial period of landscape evolution; There is a possibility of renewing damaged values. Currently, the great values of the lake are threatened by several factors such as the rapid process of wetlanding, swamping and drying in the not-so-distant future; Global climate warming and drying; Taking lake water for irrigation and electricity production; Damage and pollution of the landscape and the lake from human activity such as cutting of forests, overgrazing, fires, use of pesticides and herbicides and possible tourist overload

At the end of the work, based on history, archaeological data and findings, and field measurements, the HDZA Project built a complete GIS Map of the Cultural and Natural Heritage of the village of Lubonjë, Korçë (**Figure 5**).

6. WebGIS "Lubonja Turistike" in the promotion of cultural and natural heritage

The village of Lubonje has many areas with significant archaeological potential, cult objects and cultural and natural monuments, along with wonderful cuisine and rich natural landscapes, including forests and fields. These attributes constitute a substantial untapped tourist potential. Additionally, near the village lies the abandoned Rehova copper mine, which serves as an industrial tourism attraction. The landscape and terrain offer excellent conditions for developing activities such as mountain climbing, biking, motorcycling and driving sports cars. In this context, the WEB GIS "Touristic Lubonja" application was conceived and developed with the aim of promoting cultural and natural tourism in the Lubonja area. The project focused on integrating GIS with advanced data collection methods, such as drones, LiDAR and laser scanning, which together provide unparalleled capabilities for analyzing and safeguarding cultural assets [26].

6.1 Methodology

This site was created by following two main steps: (1) Gathering information about the current state of the village using satellite images and GIS processing; (2) Visualization of the village in both 2D and 3D. The construction of the Web GIS page utilized the ESRI platform and included the following steps: (a) Logging into the portal; (b) Loading data obtained from layers built and pre-processed in ArcGIS Pro software; (c) Ensuring layer privacy; to make the layers publicly accessible outside the organization; (d) Creation of the WEB Map; (e) Building the 3D Scene; (f) Developing the Application. WEB AppBuilder allows for the creation of an application designed according to our desired functionalities and user interests.

6.2 Creation of the WEB page

This was made possible through ArcGIS Story Maps. The WEB Page "Touristic Lubonja" will comprise all scientific information and discoveries made during the field study, integrating maps, 3D scenes and applications within the site. Additionally, it will include information about the area, its history and suggestions such as guides or descriptions of points of interest with tourist potential. The Web GIS platform is designed to be accessible on both desktop and mobile devices [27]. Access to the Map, Scene and WEB Application can be achieved by clicking the provided link or scanning the QR code. The ArcGIS Story page "Touristic Lubonja" can be accessed at the following link: https://storymaps.arcgis.com/stories/43523107564d461fbb8ea118df18efd3

7. Conclusions and recommendations

7.1 Conclusions

Based on the historical context, conversations with the community, field surveys and measurements, analysis and data processing, the project reaches the following conclusions: (1) The village of Lubonjë, Korçë, is rich with an early and unique cultural heritage. The project identified 11 areas with high archaeological potential, the most important of which is the Castle of Lubonja c. VIII-VII BC and centuries fifth to sixth AD; Kuqeshi, residence of the period c. XIII-IX BC and V-VI BC; Rrobas; Guri i Shqipes; Fusha e Lubonjës. The old historical roads and bridges are still unstudied and not mapped enough. (2) The cult objects in Lubonja, especially the Church "Dormition of Virgin" founded in the eleventh century AD, which holds one of the oldest and unique wall paintings, constitute an extraordinary scientific value that must be preserved and promoted. (3) Rich natural heritage such as centuries-old oaks and pines are remnants of former oak and conifer forests. They have great scientific, historical, didactic, cultural and religious values and are part of the spiritual world of the inhabitants. (4) The Lower Glacial Lake of Mali i Zi, completely natural, as a remnant of the last glacial period, stands out for the great purity of the water, the unique and diverse living world of flora and fauna, perhaps with endemic organisms; for its great scientific importance (geological, geographical, hydrological, biological, ecological, etc.). This lake is living in the last phase, that of wetlanding and swamping, accelerated by recent human activity.

7.2 Recommendations

Based on the above conclusions and in order to preserve and promote the cultural and natural heritage of Lubonja and the surrounding areas, the Project recommends: (1) To draw up a Regional Plan for the management and promotion of the cultural and natural heritage of the area. (2) To continue the in-depth study and survey of areas with archaeological potential such as Lubonja Castle, Kuqeshi, Guri i Shqipes, Rrobasi and Fusha e Lubonjës "Kisha Fuçinë". (3) To propose the Castle of Lubonja and the old bridges of Lubonjë, Vithkuq and Leshnjë to receive the Status of Cultural Monument, while more in-depth studies should be done for the historical roads. (4) The project strongly recommends the use of advanced GIS, DRON, Laser Scanner and LIDAR and WebGIS technologies in the mapping, documentation and promotion of Cultural and Natural Heritage at the national level. (5) In order to preserve the

Church of "The Dormition of Virgin" and promote the heritage, the project recommends the drafting of a complete Project for the Revitalization of the Church and the construction of a Museum of Cultural and Natural Heritage within the Church. (6) The project proposes that centuries-old oaks of churches; The pines near the tower of the Lubonja Castle and the lower glacial lake of Mali i Zi are given the status of Natural Monument and Protected Areas. (7) In order to preserve the lake, the main urgent measure that is recommended is the closing of the discharge channel. (8) The project recommends sensibilization of the public about the many values of the cultural and natural heritage of the area.

Acknowledgements

This publication has been made possible with the financial support of the National Agency for Scientific Research and Innovation (AKKSHI). Its content is the responsibility of the authors; the opinion expressed in it is not necessarily the opinion of AKKSHI.

The publication is carried out within the Project HDZA "*Mapping and documentation of archaeological areas using advanced GS Technology, Drone, Lidar and Laser Scanner*". The project was carried out by Polytechnic University of Tirana, Faculty of Geology and Mining; Academy of Sciences of Albania, Commission of History and Archeology and Institute of Archeology; Faculty of History and Philology; Regional Directorate of Cultural Heritage, Korça; National Museum of Medieval Art, Korça; Lubonja village. Students contributed to the realization of the study: Fjona Alikollari, Renata Myftari, Valter Aga, Denis Nika, Elmira Alushaj, Xhesika Buraku, Xhesian Toska and Xhejms Krasniqi.

Conflict of interest

The authors declare no conflict of interest.

Gratitude

The authors of the study thank AKKSHI for supporting the Project; Polytechnic University of Tirana, Faculty of Geology and Mining; Academy of Sciences of Albania, Commission of History and Archeology; Institute of Archeology; Faculty of History and Philology; The Regional Directorate of Cultural Heritage Korçë, the National Museum of Medieval Art Korçë and the central and local authorities for their support. In particular, the authors thank the community of the village of Lubonjë for their valuable contribution and their warm welcome, particularly Albert Hime, Xhemal Velikollari, Myftar Xhakollari, Arshin Serani, Ilirjan Beqollari, Servet Hoxha, Sazan Liço, Klaudio Serani, Ermir Hoxha, Metjol Hime, Oljan Ibro and Ardit Hoxha.

Author details

Edmond Hoxha[1*], Neritan Ceka[2], Skënder Aliu[3], Perikli Qirjazi[2], Jeton Pekmezi[1], Elio Hobdari[4], Elva Leka[1], Emiliano Qose[5], Erland Alla[6], Spartak Kuçaj[1], Era Xhaferaj[7], Fjoralba Prifti[8], Genti Proko[8] and Jani Zoica[9]

1 Faculty of Geology and Mine, Polytechnic University of Tirana, Tirana, Albania

2 Academy of Sciences of Albania, Tirana, Albania

3 Independent Archaeologist, Tirana, Albania

4 Institute of Archaeology, Academy of Sciences of Albania, Albania

5 Alb Matrix Group Ltd, Albania

6 State Authority for Geospatial Information, ASIG, Albania

7 National Institute of Cultural Heritage, Albania

8 National Museum of Medieval Art, Albania

9 Regional Directorate of Cultural Heritage, Korçë, Albania

*Address all correspondence to: ehoxha63@gmail.com

References

[1] Xhaxho M. Lubonja në të vërtetat historike. Monografi, Kotti, Korçë 2004. fq. 25, 26

[2] Xhaxho M. Lubonja në të vërtetat historike. Monografi, Kotti, Korçë 2004. fq. 22-23

[3] Conolly J, Lake M. Geographical Information Systems in Archaeology. Cambridge: Cambridge University Press; 2006

[4] Marwick B, Hiscock P, Sullivan M, Hughes P. Landform boundary effects on Holocene forager landscape use in arid South Australia. Journal of Archaeological Science: Reports. 2017;**19**:864-874. DOI: 10.1016/j.jasrep.2017.07.004. S2CID 134572456

[5] Flygues. Lidar & archaeology [Internet]. 2024. Available from: https://flyguys.com/lidar-technology-helps-archaeology-industry/ [Accessed: 2025 January 4]

[6] LinkedIn. How can laser scanning complement other methods of archaeological survey and excavation? [Internet]. 2025. Available from: https://www.linkedin.com/advice/3/how-can-laser-scanning-complement-other-methods [Accessed: 2025 January 4]

[7] Universiteti i Tiranës. Burime të zgjedhura për Historinë e Shqipërisë, Tiranë 1962. f. 30-31

[8] Aliu S. Kolonja – Studime arkeologjike. Fq.87. Kotti 2020. ISBN 978-9928-299-26-0

[9] Xhaxho M. Lubonja në të vërtetat historike. Monografi, Kotti, Korçë 2004. fq.17, 19, 23

[10] Xhaxho M. Lubonja në të vërtetat historike. Monografi, Kotti, Korçë 2004. fq.28

[11] Xhaxho M. Lubonja në të vërtetat historike. Monografi, Kotti, Korçë 2004. fq.17

[12] Lamarck J-P. 1801. Systême des animaux sans vertèbres; ou, Tableau général des classes, des classes, des orres et des genres de ces animaux. Deterville, Paris. 1801. viii + 432 pp. Coral "Cyclolites ellipticus". Identified by Dr. Andrea Uta polytechnic University of Tirana. Institute of Geosciences. Geology Department. 2024

[13] Lera P. Vështrim arkeologjik në krahinën e Vithkuqit. 1980. Gazeta "Përpara" 8.08.1980

[14] Xhaxho M. Lubonja në të vërtetat historike. Monografi, Kotti, Korçë 2004. fq.27

[15] Karaiskaj G, Lera P. Fortifikimet e periudhës së parë të Hekurit në rrethin e Korçës. Kuvend i Parë i Studimeve Ilire. 1974;**V**(II):263-290

[16] Aliu S. Kolonja – Studime arkeologjike. Fq.86. Kotti 2020. ISBN 978-9928-299-26-0

[17] Karaiskaj Gj. 5000 vjet fortifikime. 1981, f. 23

[18] Xhaxho M. Lubonja në të vërtetat historike. Monografi, Kotti, Korçë 2004. fq.17, 26

[19] Xhaxho M. Lubonja në të vërtetat historike. Monografi, Kotti, Korçë 2004. fq.30

[20] Popa Th. Mbishkrime të Kishave të Shqipërisë. Tiranë 1998, mbishkrimi nr.398, fq.190

[21] Popa Th. Miniatura dhe piktura mesjetare në Shqipëri. 4. Piktura më e vjetër e Kishës së Shën Marisë në Lubonjë të Kolonjës". fq.110

[22] Lozanova R. СРЕДНОВЕКОВНИ СТЕНОПИСИ В ЦЪРКВАТА В СЕЛО ЛЮБОНЙЕ (АЛБАНИЯ) "Sofia University", "St. Kliment Ohridski" Center for Slavic-Byzantine Studies "Ivan Duychev", Sofia. 2003. ISSN-1311-784X; Annuaire de l'Université de Sofia St. Kliment Ohridski. Centre de recherches slavo-byzantines Ivan Dujčev, Tome 92 (11), 303-305

[23] Qirjazi P. Veçori gjeomorfologjike të pellgut të Kolonjës. Përmbledhje Studimesh, Nr. 4/1980

[24] Qiriazi P. Morfologjia dhe Morfogjeneza e Gropave Juglindore dhe e maleve përreth tyre. Universiteti i Tiranës, Fakulteti i Historisë dhe i Filologjisë. 1985

[25] Qirjazi P. Trashëgimia natyrore e Shqipërisë, botim i ASHSH. 2019

[26] Yao Y, Wang X, Luo L, Wan H, Ren H. An overview of GIS-RS applications for archaeological and cultural heritage under the DBAR-heritage mission. Remote Sensing. 2023;**15**(24):5766

[27] Vasilev J, Petrov P, Jordanov J. A practical approach of data visualization from geographic information systems by using mobile technologies. International Journal of Interactive Mobile Technologies (IJIM). 2024;**18**(03):4-15. DOI: 10.3991/ijim.v18i03.46655

Chapter 6

Monitoring Shrub Cover Change across a Disturbed Semi-Arid Island Landscape with Repetitive Aerial Imaging

Douglas Stow, Kelsey Warkentin, Kellie Uyeda, John O'Leary, Julie Lambert, Andrew Loerch, Hsiao-chien Shih, Lloyd Coulter, Leslie Bolick and Kimberly O'Connor

Abstract

Monitoring shrub cover change with very high resolution (VHR), multi-temporal imaging supports adaptive management of disturbed semi-arid island landscapes where native and non-native species of shrubs and herbs commonly compete. We address the potential of aerial orthoimagery time series for shrub cover monitoring for San Clemente Island, California, USA by: (1) analyzing shrub cover change using orthoimages for the period 2010 to 2018, (2) determining the utility of normalized digital surface models (nDSMs) derived from highly overlapping VHR imagery, and (3) designing an end-to-end aerial shrub cover monitoring system. Extant orthoimage data sets tend to have inconsistent times of image capture and spatial resolutions, which limits reliable monitoring of shrub cover changes. Multi-temporal orthoimages and nDSMs derived from Structure from Motion (SfM) photogrammetry are useful for shrub cover monitoring. The key to a successful aerial imaging system for monitoring shrub changes is replicating image capture and processing.

Keywords: shrub cover, vegetation change, aerial imaging, vegetation monitoring, Mediterranean-type ecosystems, southern California vegetation, shrubland

1. Introduction

For many heavily disturbed semi-arid landscapes around the world, vegetation may transition from native shrubland to a non-native grassland community type [1]. This is particularly the case for the Channel Islands in the Southern California Bight off the coast of California, USA. Following cessation of one or more disturbance agents and/or upon implementation of restoration or other mitigation measures, recovery may begin to occur with vegetation transitioning back to community composition before heightened disturbance(s) occurred [2]. Such transitions may also limit other ecosystem services associated with wildfire, soil stabilization and carbon sequestration. Monitoring response and recovery to disturbance may be of interest to resource managers and/or conservation biologists working in these semi-arid ecosystems, particularly when native vegetation is an important element of wildlife habitat requirements [2, 3].

Spatially comprehensive monitoring of heavily disturbed and recovering semi-arid island landscapes can only be achieved through some form of multi-temporal remote sensing. Research over the past decade has shown that tracking temporal changes in fractional cover of growth form types is a promising approach to monitoring semi-arid landscapes [4, 5]. Changes in vegetation growth form cover yield stronger and earlier signals related to ecosystem transitions and response/recovery associated with disturbances in semi-arid landscapes than species- or community-level changes [6]. Monitoring growth form cover changes in these landscapes primarily involves detecting and preferably quantifying transitions from herbaceous to shrub plant forms, or vice-versa. An example of such a transition is a chaparral stand of shrubs that burns more than once in sequential fires at a relatively short return interval, does not recover and is mostly replaced by non-native grasses, which establish following seed dispersal from nearby grass patches [7].

Very high (spatial) resolution (VHR) imagery is particularly useful for quantifying finer-scale growth form transitions, meaning that information on vegetation community decline or recovery can be gleaned sooner, and with greater potential for successful mitigation or adaptation by resource managers. However, monitoring based on multi-temporal VHR imagery may be limited by the lack of temporally extensive (limited inter-annual coverage) and consistent (similar intra-annual conditions) image acquisition and archive collections. Affordability is also an issue since most organizations charged with biological monitoring are government or non-profit organizations. Sources of affordable VHR multi-temporal imagery are limited to certain types of aerial orthoimagery, such as from governmental subsidized programs and some commercial entities, some of which may have sufficient spatial resolution but are typically not captured at sufficient temporal frequency and/or tend to be prohibitively expensive.

The National Agricultural Imagery Program (NAIP) of the United States of America (USA) is the latest of several aerial orthoimagery production and dissemination programs for the US, starting in the mid-1990s and is essentially a free source of VHR visible and near infrared (V/NIR) imagery for vegetation monitoring [8]. Geometrically processed visible and near infrared waveband images are available for most of the US on a bi-annual basis, with a ground sampling distance (GSD) of 1.0 m spatial resolution and since 2016, GSD = 0.6 m. The seasonal date of acquisition can vary, though generally collected in the late spring or early summer for the southern California study area associated with this article. While NAIP and similar orthoimage programs provide freely available VHR orthoimagery for the past few decades in support of retrospective studies, the lack of consistency in imaging characteristics over time may reduce the effectiveness of vegetation change monitoring.

The advancement and integration of several components associated with aerial orthoimage production presents an opportunity to conduct affordable, consistent and reliable monitoring of semi-arid landscapes based on VHR multi-temporal imagery. These components include off-the-shelf digital cameras with image arrays in the 20–100 megapixel range, location-based aircraft navigation and camera triggering systems based on real-time kinematic (RTK), global navigation satellite systems (GNSS), and Structure from Motion (SfM) photogrammetric processing software. Two or more digital cameras, with at least one modified to be NIR sensitive, may be mounted and synched to acquire V/NIR imagery. Free or inexpensive flight planning software allows for flight planning and image capture with a high degree of image overlap, by synchronized camera triggering based on aircraft positions derived from the RTK-GNSS system. Image station positions in the sky can be matched over time using the same flight planning, software, and navigation and triggering system. Automatic image registration software enables images from multiple cameras and for matched imaging stations over time to be precisely registered, yielding a multispectral, multitemporal layer stack. Layer stacks for each camera station are subject to SfM processing to automatically generate a digital surface model (DSM), which provides the basis for orthorectification of relief displacement, such that an orthorectified, seamless image mosaic is generated for each image date. Object-based image analysis software is then applied to these VHR orthoimage mosaics to generate maps of growth form types over time. These procedures and capabilities are elaborated upon throughout this chapter.

For this chapter, we build on the results from past research and associated data sets to assess the potential and prescribe specifications for an operational aerial image-based monitoring system for disturbed semi-arid landscapes. In particular, we build on Uyeda et al. [9] and Warkentin et al. [10], which explored community-level and growth form cover mapping based on VHR aerial orthoimagery for San Clemente Island (SCI), off the coast of southern California, USA. Disturbance from grazing and military activities on the island converted what was mostly coastal sage scrub shrublands to a non-native grassland, which is now beginning to recover with the cessation of grazing and in many areas, more limited mechanical disturbance from military exercises.

Uyeda et al. [9] demonstrated that vegetation communities could be delineated and classified at the alliance level with a high degree of automation and reliability using multi-scale OBIA classification. A finer-scale growth form classification with shrubs classified at the species level, served as input to a coarser scale segmentation and classification of community sub-units, based on fractional cover rules. The sub-units were merged into single alliance-level community units to produce the final vegetation community map.

Warkentin et al. [10] further explored growth form-level classification and mapping with aerial orthoimage data sets of varying spatial resolutions based on imagery captured over a 9-year period. Data sets included NAIP and finer spatial resolution (0.15 m) orthoimagery created from digital aerial multispectral images captured by Near Earth Observation Systems (NEOS), LTD. Both pixel-based and OBIA classification approaches were implemented. The authors found that the finest spatial resolution (0.15 m) orthoimage sets with OBIA classification yielded growth form maps with the most accurate shrub cover estimates. Both Snavely et al. [11] and Warkentin et al. [10] found that inclusion of a LiDAR-derived normalized digital surface model (nDSM) to represent canopy heights as part of the image classification made a significant and important improvement in estimating and mapping shrub cover.

The first objective and component of this chapter is to utilize a set of multitemporal orthoimages for the period 2010 to 2018 compiled and classified as part of the Warkentin et al. [10] study to evaluate the potential and limitations of such data for monitoring changes in shrub cover for SCI. The second objective and component is to explore the utility of normalized digital surface models (nDSMs) derived from highly overlapping VHR image sets to generate canopy height raster data sets to be used as input to image classifiers for mapping shrub cover distributions and changes. The third objective is to design and specify an end-to-end aerial imaging system that best meets requirements for monitoring shrub cover change.

2. Potential of shrub cover monitoring based on orthoimagery archive

The reliability of quantitative estimates and maps of vegetation cover change depends on the accuracy and temporal stability of these maps [12]. Separating classification errors, artifacts and uncertainty associated with misregistration, persistent cover, and actual cover changes of growth forms (i.e., expansion and decline) from image-derived map series is challenging. Variable dates/seasons of acquisition, image types and characteristics, and image processing approaches make monitoring shrub changes even more challenging.

We analyzed the potential for quantifying shrub cover dynamics over an 8-year period (2010–2017) for SCI based on shrub cover maps derived through classification of aerial orthoimagery, as reported in Warkentin et al. [10]. We hypothesized that shrub cover had increased slightly over the past decade, with some shrub recovery following stoppage of past grazing and greater localization of mechanical disturbances associated with military training within what had become a predominantly non-native grassland across most of SCI. An assumption of slight expansion of shrub cover is based mostly on anecdotal and qualitative evidence, as no permanent field plot or transect data sets appear to be available to document grassland-shrubland change over the past decade. Relatively rapid expansion of *Baccharis pilularis* could be expected, as this shrub species can colonize within a few years after disturbance [13]. However, the period 2011 through 2017 is characterized by predominantly drought conditions, which may have reduced shrub vigor and possibly cover for some SCI locations.

2.1 Data and methods

Our analysis of apparent trends in shrub cover distributions is based on two 2 km × 2 km areas of interest (AOIs) within SCI that were the major study sites for the Warkentin et al. [10] study. These two AOIs were selected based on having a composition of shrub and non-shrub growth forms and range of shrub cover representative of SCI in general. Growth form-level classification and mapping of 2010–2018 aerial orthoimage data sets (**Table 1**) were performed following methods described in Warkentin et al. [10]. Both pixel- and object-based image classification approaches were implemented, depending on the spatial resolution of the orthoimages. An nDSM derived from airborne LiDAR data collected in 2014 was included as input for the image classification of shrub cover. Shrub cover fraction maps were generated from the shrub cover classification based on a grid of 40 m × 40 m cells.

Within the two AOIs, 25 randomly selected plots, with dimensions of 40 m × 40 m, were used to estimate fractional growth form cover as reference data for assessing the accuracy of shrub cover estimates from semi-automated image classification. Reference shrub cover estimates for sample plots, used for assessing accuracy, were based on visual interpretation of RGB images captured in 2018 and having 0.08 m spatial resolution (**Table 1**). Image interpretations were calibrated by direct field observations at 100 regularly spaced grid points for five of the grid cells.

2.2 Results

Variation in apparent shrub cover over the study period is shown in **Figure 1**, with mean shrub cover estimates from 25 40 m × 40 m reference data plots grids and total mapped shrub cover for the entire AOI for each of the 10 shrub cover map. Since the

Image	Date	Format	Spatial Resolution
NAIP 2010	May 2010	4-band RGB, NIR	1 m
NAIP 2012	April 2012	4-band RGB, NIR	1 m
NAIP 2014	July 2014	4-band RGB, NIR	1 m
NEOS 2015	Nov. 2015	4-band RGB, NIR	0.15 m
NAIP 2016	July 2016	4-band RGB, NIR	0.60 m
NEOS 2017	Aug. 2017	4-band RGB, NIR	0.13 m
NEOS 2018	Nov. 2018	4-band RGB, NIR	0.08 m

NAIP—National Agricultural Imagery Program; NEOS—Near-Earth Observation Systems; RGB—red, green, blue; NIR—near infrared.

Table 1.
Image datasets used for the classification of shrub and non-shrub cover, derivation of percentage shrub cover estimates (2010–2017), and generation of reference data for accuracy assessment (2018).

field-based reference grids had varying amounts of shrub cover, the standard deviations tend to be large, with an average of 15.38%.

Early (2010, 2012, 2014) NAIP image pixel classification products overestimate shrub cover (i.e., shrub cover mapped in the AOI exceeds shrub cover mapped in the reference data), which may be due in part to their relatively coarse (1 m) spatial resolution. Overestimation is more pronounced in 2010 and 2012 imagery, which was captured in spring months when herbaceous and shrub vegetation are both green. 2014 imagery was captured in July when herbaceous vegetation typically starts senescing and 2014 was a drought year, so peak green biomass of both herbs and shrubs was likely reduced, resulting in relatively low reference grid shrub cover estimates.

All three classification methods using the 2016 NAIP higher resolution (0.6 m) imagery captured in July show mapped shrub cover estimates within 1% of reference grids and the overall reference average. The 2016 NAIP OBIA with nDSM classification map shows shrub cover estimates only 0.15% over the reference shrub cover average, the closest of all maps.

Aerial orthoimages from later in the study period, were generated from NEOS imagery (2015–2017) and have higher spatial resolution (0.08–0.15 m). These images were captured in summer or fall when herbaceous and shrub growth forms are strongly differentiated, yielding more conservative shrub cover estimates. Thus, when tracking apparent change using these datasets, shrub cover appears to have declined for many parts of the study areas.

Based on the percent overall cover estimates for the various image products relative to the reference data, the two products that were the most accurate for the site-specific accuracy assessment of the 10 shrub cover maps (2015 and 2017 OBIA with nDSM) in the Warkentin et al. [10] study, exhibited the greatest difference in overall areal cover of shrubs relative to that estimated from the reference data. The 2015 and 2017 OBIA with nDSM products yield maps that depict a lower amount of shrub cover compared to those based on image inputs only. Shrub cover estimated from the reference data is based on a spatial sample of the area represented in a spatially complete (wall-to-wall) manner by the image classification shrub cover maps, which may account for this discrepancy in cover estimates.

Shrub cover estimated for the early period (2010–2014) plotted against shrub maps in the later period (2015–2017) using a 1-m common grid in **Figure 2** portrays

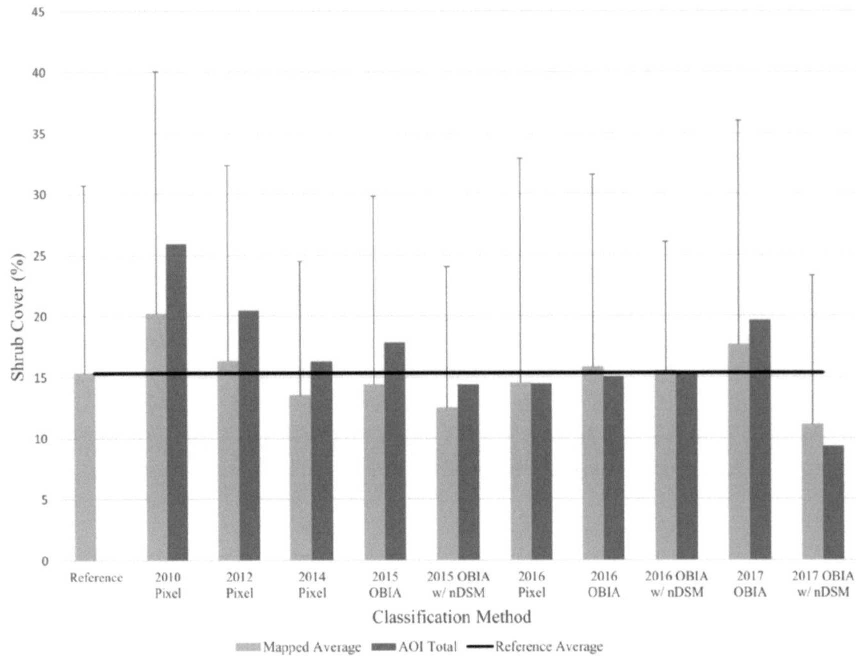

Figure 1.
Overall shrub cover (mean cover within AOIs 1 and 2; n = 25) and reference data (mean cover within reference grids; n = 100). Reference grid data shown with standard deviation "whiskers" with values ranging from 11.11% to 20.19%.

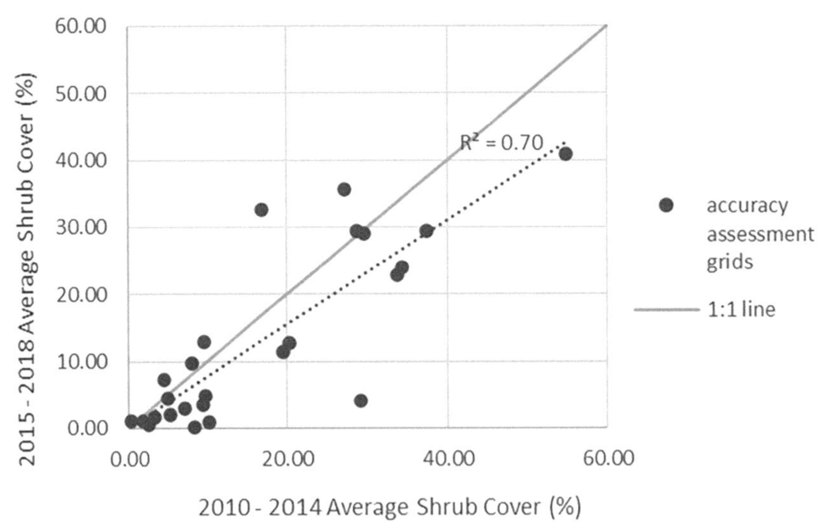

Figure 2.
Early (2010–2014 NAIP) versus late (2015–2018 NEOS) shrub cover averages for the 40 × 40 m grids for AOIs 1 and 2. Each point represents an accuracy assessment grid. Points that fall below the 1:1 line represent grids where more shrub was classified in the early NAIP imagery. Points that fall above the 1:1 line represent grids where more shrub was classified in the later NEOS imagery. Points on the 1:1 line represent grids that had the same amount of shrub classified over the length of the study period. The best fit (dotted) line for accuracy assessment grids shows the relationship between mapped cover in 2010–2014 vs. 2015–2018 datasets.

a trend of shrub loss, again likely due to the over-classification of shrubs using the coarser spatial resolution, spring season NAIP imagery (**Figure 2**). The $R^2 = 0.70$ and the least squares line falls below the 1:1 line, suggesting that most of the grids were mapped with more shrub cover in the early period than the late period. Of the 25 reference grids sampled within AOIs 1 and 2, only seven show an apparent increase in shrub cover over the duration of the study period. (AOI 3 was not analyzed for apparent shrub change since only the imagery from 2015 to 2018 was classified.) The average difference in shrub cover between the earlier and later periods is −3.65%, indicating an apparent overall reduction of shrub cover.

By comparing shrub distributions of selected maps representing the beginning of the study period (2010 and 2012 NAIP classification products) to those from the end of the period (2015 and 2017 NEOS classification products), we attempted to detect actual shrub expansion in likely locations (see **Figure 3**). Summing percent cover of the different classification products at a common 1 m grid level reduces effects of misregistration and classification errors. The 40 m fractional cover product shows a greater apparent expansion of shrubs, while the 20 m fractional cover indicates finer-scale patterns of apparent growth within those larger areas.

Shrubs that were mapped with the higher spatial resolution NEOS imagery might be considered to represent actual shrub growth or expansion, since the classification of earlier-period, coarser spatial resolution and early-season imagery tended to over-map shrubs. Thus, we performed visual image interpretation of the 2010 NAIP, 2017 NEOS, and 2018 high-resolution reference data imagery to assess whether shrub growth/expansion represented on image classification products actually occurred. An example of results from this assessment are depicted for sampling Grid 3 in **Figure 4**. This grid was selected because it has the highest apparent

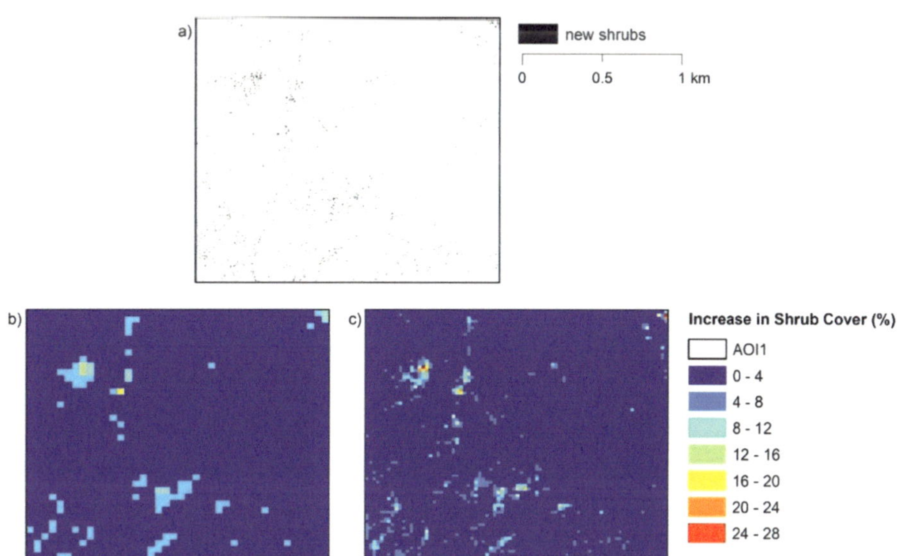

Figure 3.
Shrub change between the early NAIP and later NEOS imagery in a portion of AOI 1. (a) Newly mapped shrubs that were classified in the 2015 and 2017 NEOS imagery but not classified in the 2010 and 2012 NAIP imagery; (b) 40 m fractional cover of new shrubs; (c) 20 m fractional cover of new shrubs.

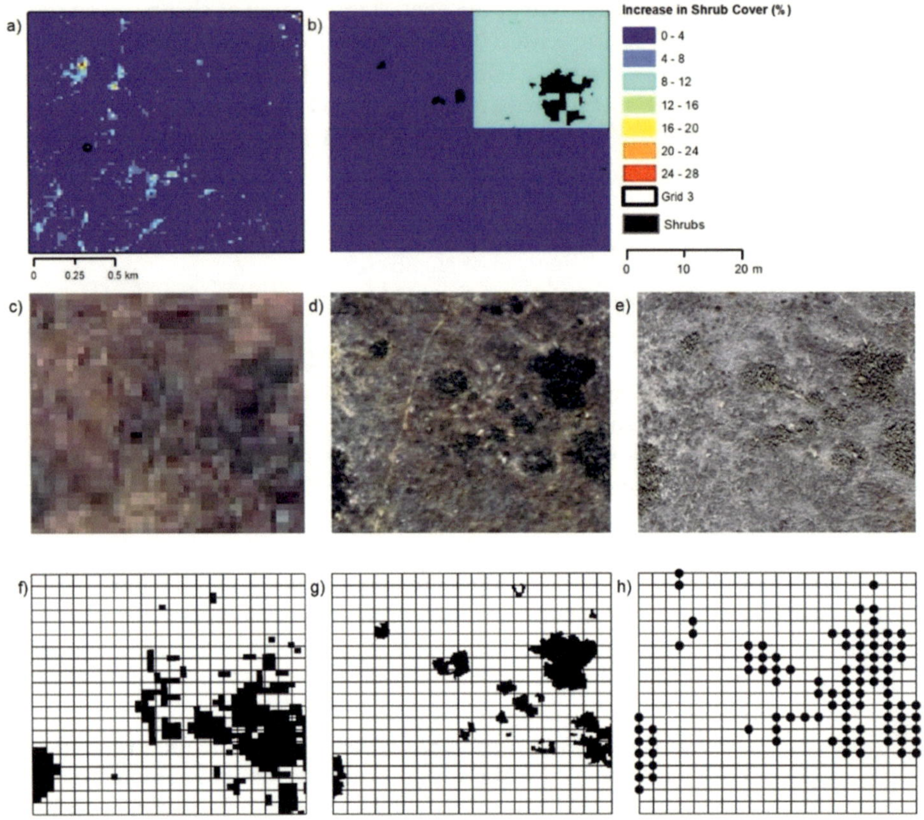

Figure 4.
Visual assessment of apparent shrub expansion and growth based on comparison of 2017 NEOS imagery and 2010 NAIP imagery for a portion of field sampling grid 3. (a) 20 m fractional cover grid with newly mapped shrubs from the 2017 NEOS imagery; (b) closeup of accuracy assessment grid 3 with the new shrub classification overlay; (c) 2010 NAIP imagery; (d) 2017 NEOS imagery; (e) the accuracy assessment imagery; (f) 2010 pixel classification; (g) 2017 OBIA with nDSM classification; (h) accuracy assessment image interpretation shrub classification.

expansion of shrub cover. Between the early NAIP and later NEOS periods, shrub cover estimates for Grid 3 at the 20 m level exhibit an average increase of 4.12%. The 2017 NEOS classification matches the accuracy assessment sampling points more closely than the 2010 NAIP product (**Figure 4e, f, g**). The 2017 NEOS imagery is clear enough to discern shrubs from herbs, while the 2010 NAIP is too coarse to accurately separate shrubs from herbaceous cover, which demonstrates the likely cause of classification errors and over-estimation of shrub cover in the 2010 NAIP imagery.

Even shrub change maps derived from the most accurate maps show inconsistencies in image classification products that appear as substantial shrub changes over just 2 years. The classification products derived with the 2015 and 2017 NEOS and nDSM inputs to the OBIA classifier show substantial differences that appear as shrub decline, but more likely represent differences in imagery characteristics or classification parameters. Differences in shrub cover at the 40 m grid level between the 2015 and 2017 NEOS OBIA with nDSM products range from −21.91% (loss of shrub cover) to 41.76% (shrub growth) (**Figure 5**). The average difference between the 2 years represents a loss of 3.30% shrub cover.

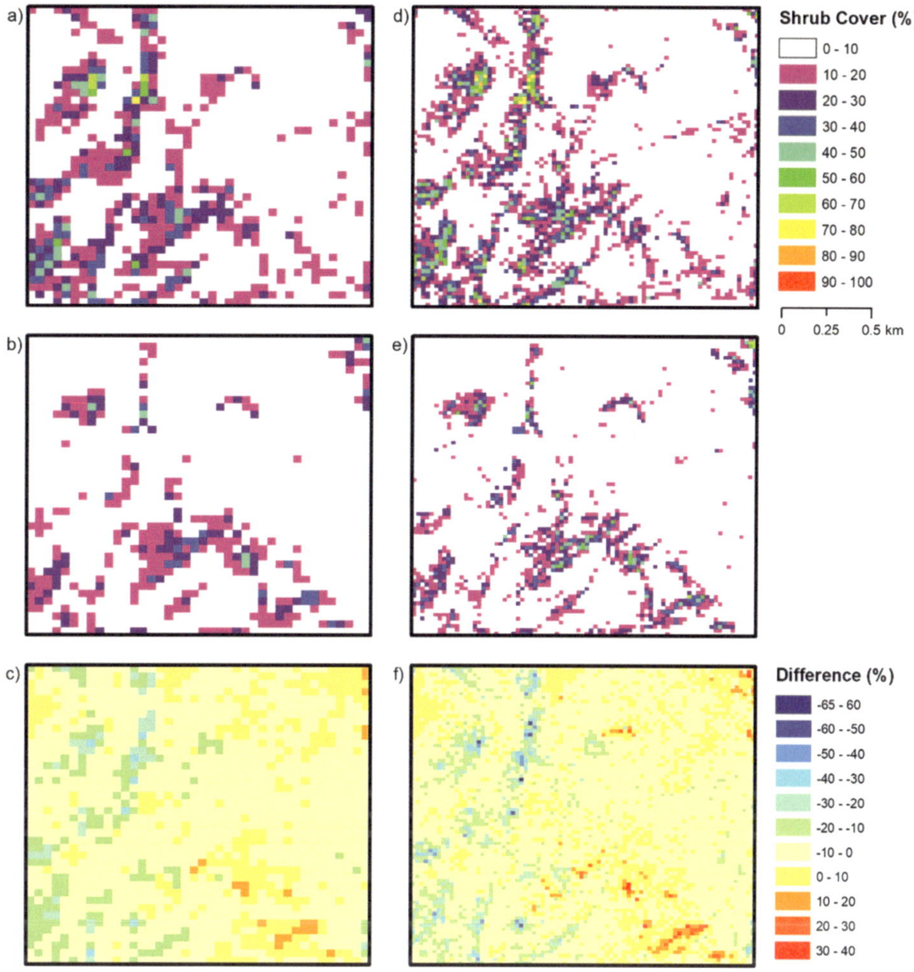

Figure 5.
Fractional cover products for OBIA with nDSM methods at 40 m grid level (column a–c) and 20 m grid level (column d–f) (a) 2015 NEOS. (b) 2017 NEOS. (c) Difference showing shrub change at the 40 m grid. (d) 2015 NEOS. (e) 2017 NEOS. (f) Difference shrub change at the 20 m grid level. The greatest decrease (loss) in shrub cover is −21.91% while the greatest increase (growth) is 41.76% with these extremes displayed only for the 20 m grid. The mean difference is −3.30%.

2.3 Summary

A clear finding from this analysis of the shrub cover map series generated by Warkentin et al. [10] is that monitoring of shrub cover changes in a severely disturbed semi-arid shrubland landscape is not consistently reliable nor accurate when based on classification of extant VHR orthoimagery. This primarily stems from inconsistencies in spatial resolution, seasonality of image capture, and geometric processing. These inconsistencies yield differences in spectral-radiometric signatures of shrub and non-shrub growth forms and land covers, and misregistration of image-derived maps, such that even shrubs that are persistent over a study period are not rendered consistently and influence changes in apparent shrub cover fractions.

The coarser spatial resolution (1 m rather than 0.15 or 0.6 m) of NAIP orthoimage data and lack of time-specific nDSM for the earlier years made accurate classification and cover estimation of shrubs challenging. We found that the 1 m image spatial resolution is too coarse to accurately resolve and identify some shrub individuals, resulting in errors and uncertainty in representing shrub cover over space and time.

The inclusion of an nDSM as an input feature to image classifiers is important for distinguishing shrubs from non-shrub land cover when classifying VHR images of SCI; this is particularly the case when images are captured at a time when the multispectral image reflectance signatures of sparse and low stature shrubs are similar to that of the predominant herbaceous vegetation cover. A single nDSM derived from airborne LiDAR data captured in 2014 was used as a feature input along with the various dates and sources of VHR orthoimages in the Warkentin et al. [10] study. This means that shrub canopy height representation is static input to image classifications spanning the entire study period. Given the importance of nDSM inputs to image classification of shrub and non-shrub categories, we explore the reliability of nDSM derived from SfM processing of VHR imagery in the second study component which follows.

3. Potential of nDSMs for enhancing shrub monitoring

We evaluated the potential for mapping vegetation canopy height distributions using nDSMs derived from SfM processing of highly overlapping VHR image frames. We compared the resulting nDSMs to one produced from airborne LiDAR data that was used in the study component above. SfM nDSMs are less expensive to obtain compared to LiDAR data, but tend to be less accurate. When processing LiDAR datasets, the ground surface can be readily determined by selecting the last return. Detecting the ground is more difficult in SfM point clouds, as only the surface of the canopy tends to be represented in point clouds generated using this technique.

3.1 Data and methods

Highly overlapping airborne digital (three visible and one NIR waveband) imagery for the two AOIs examined above provided source data for deriving DSMs and nDSMs for evaluation of canopy height mapping capabilities. Images were captured with approximately 80% forward and sideward overlap. A subset of the original imagery collection area was processed using the SfM technique due to issues with collection of GCPs that constrained the final extent.

Survey-grade global navigation satellite system (GNSS) coordinates were captured in the field to provide ground control for SfM processing. An Emlid real time kinematic GNSS unit was used to measure positional (x-y-z) coordinates to the cm-level for stable ground features that could be identified on VHR image sets. Field surveyors also captured ground-based images to document the positions and characteristics of surveyed points.

SfM processing of the VHR image sets was conducted using Agisoft Metashape software. Image frames, GNSS logs for aerial image station positions, and surveyed ground control points were input to Metashape to generate a DSM for the scene extent. The DSM was used for correcting for terrain-related parallax to create an orthoimage mosaic and to derive an nDSM. The DSM and orthoimagery were geographically referenced based on the geometric and positional calibration provided by the surveyed ground control points.

Normalized digital surface models (nDSMs) from SfM-derived DSMs were generated using the SfM-derived point cloud. Points were classified as vegetation or ground using NDVI thresholds and contextual information as part of an iterative process. In the initial step, ground points were classified using automatic algorithms within Global Mapper software. Next, a NDVI threshold of 0.2 was selected to initially identify vegetation. To more fully capture the extent of shrubs, a lower threshold of 0.1 was applied to classify points within 1.5 m of vegetation points. Finally, all ground points within 0.5 m of shrubs were removed. This final step ensured that erroneous ground points were not located within shrub canopies. These classified points were used to generate both a DEM from the ground points and a DSM from the highest elevation ground and non-ground points using a raster surfacing routine. The raster DEM was then subtracted from the raster DSM to produce the nDSM, or canopy height.

The SfM-generated nDSM was compared to the LiDAR-based nDSM. Although the point density of the SfM dataset was higher than that of the LiDAR point cloud, we used a spatial resolution of 0.5 m for the raster outputs of both products. To compare spatial patterns, we also generated a 40 m grid and calculated the area covered by vegetation measuring at least 0.2 m tall based on both the LiDAR and SfM nDSMs. The average height of *Baccharis pilularis* shrubs was also estimated for ten 2 m × 2 m areas in the LiDAR and SfM nDSM products. The SfM nDSM was registered to the LiDAR nDSM to ensure the same areas were being compared in both products, and shrubs were selected based on the original Level 1 shrub mapping described in Uyeda et al. [9].

3.2 Results

The final nDSM product shows clear delineation of shrub patterns (**Figure 6**). The overall pattern of the SfM canopy height layer is similar to that of the 2014 LiDAR-based canopy height layer (**Figure 7**). Likewise, the pattern of shrub cover is similar in both the LiDAR and SfM products (**Figure 7**).

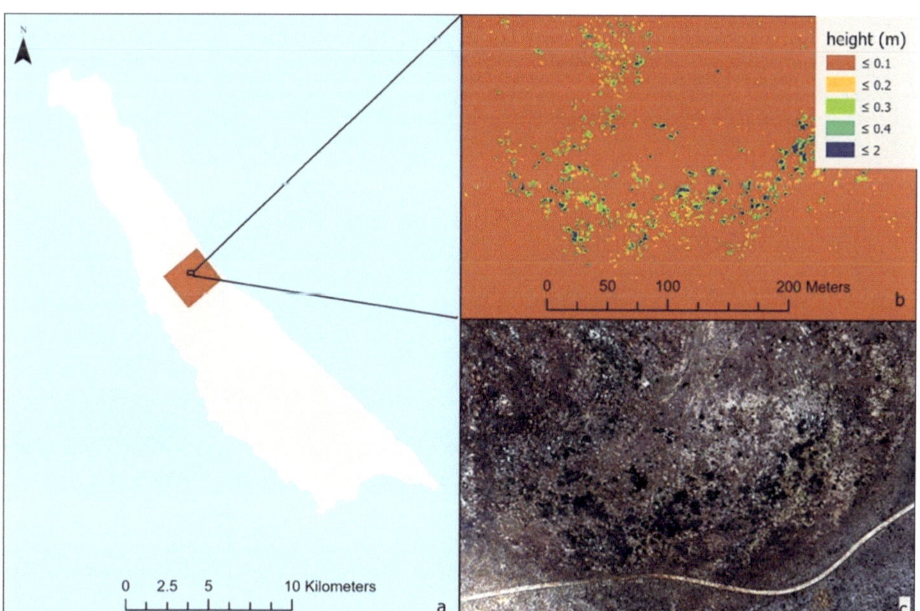

Figure 6.
Overview of SCI showing (a) SfM nDSM processing extent, (b) inset of SfM nDSM, (c) inset of corresponding NEOS 2017 image.

Although the overall spatial pattern of shrub detection appears similar in both the LiDAR and SfM nDSM products, the height detected in the SfM product tended to be lower, as shown in the comparison of *Baccharis pilularis* individuals measured

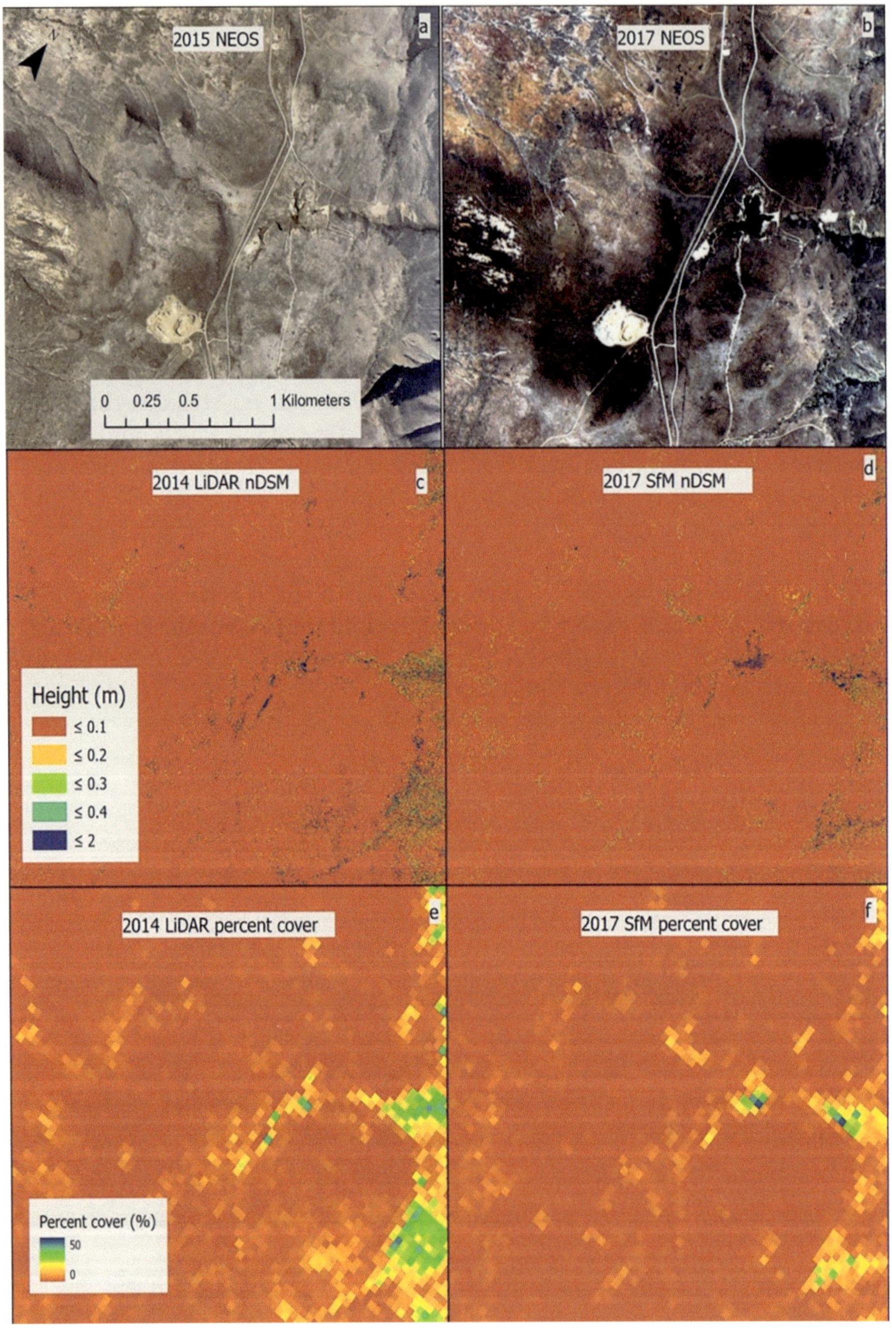

Figure 7.
Comparison of canopy height products, showing (a) 2015 NEOS imagery, (b) 2017 NEOS imagery, (c) canopy height of 2014 LiDAR product, (d) canopy height of 2017 NEOS Structure from Motion (SfM) product, (e) percent cover of all pixels with height greater than 0.2 m based on LiDAR processing, (f) percent cover of all pixels with height greater than 0.2 m based on SfM processing.

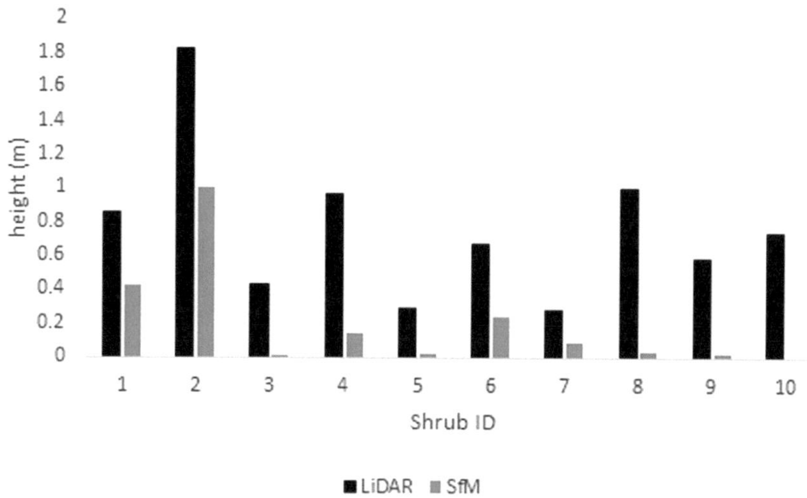

Figure 8.
Shrub height based on LiDAR and SfM for 10 sampled Baccharis pilularis *individuals.*

with both techniques (**Figure 8**). It is possible that the actual heights of shrubs changed somewhat from 2014 to 2017, although likely not to the extent observed here. It appears that the SfM nDSM did not reliably estimate the full height of canopies compared to the LiDAR nDSM.

The SfM nDSM product reliably represented the change in patterns of growth of the herb *Salsola* sp. (Tumbleweed, or Russia thistle). Unlike shrubs, this rapidly growing annual plant does not persist for long in a particular location. Thus, distinct spatial patterns in growth are apparent when comparing multi-temporal nDSMs.

3.3 Summary

Structure from Motion processing shows potential for shrub canopy representation and identification for open shrublands. The spatial resolution of the canopy height layer is improved relative to the LiDAR canopy height, allowing for the detection of small shrubs and clear delineation of shrub boundaries. While the nDSM product exhibited vertical accuracy levels that were less than nDSM generated from LiDAR data, the primary motivation behind this processing was detecting shrub height signals that could be exploited for classifying and mapping shrubs relative to low-lying herbaceous vegetation. The SfM canopy height layer clearly distinguishes from a background herbaceous vegetation (mostly non-native grasses), increasing shrub identification accuracy [10]. Additional research and development is needed to further improve the vertical accuracy estimation of canopy heights. In the future, repeated monitoring incorporating temporal-specific nDSMs should provide a low-cost and efficient means for monitoring shrub expansion and growth, as described in Component 3 below.

4. Proposed shrub cover monitoring system

The third component of this chapter pertains to specifying requirements for an operational end-to-end aerial imaging system for monitoring interannual changes in shrub cover for SCI and other recovering areas composed primarily of coastal sage scrub and similar shrubland communities. Spatially-explicit monitoring of shrub recovery following severe disturbance such as from grazing, other mechanical disturbances, drought and severe and/or frequent fire is important for supporting habitat management and ecological restoration [14]. Similarly, monitoring shrub cover may also provide information on locations of shrubland degradation for which management action may need to be intensified. Specifications provided in this section are informed by our empirical analyses of shrub classification and cover estimation based on extant aerial orthoimage datasets covering SCI, even though the production and characteristics of these orthoimage data sets may not be sufficiently consistent and suitable for reliable reconstruction of shrub cover change for the 2010 through 2017 study period.

What was verified through our analyses above is that successful monitoring of vegetation and land cover using multi-temporal remote sensing approaches is best achieved by replicating image collection and processing parameters and procedures over time [15, 16]. Such replication yields multi-temporal image data sets that are more spectrally, radiometrically (i.e., image brightness), spatially, and geometrically consistent, than image sets generated in an *ad hoc* and temporally different manner. Replicating dates and times of image collection aids in minimizing differences in solar illumination and shading, vegetation phenology, and soil moisture properties between years. Such temporal consistency means that signals of vegetation change, primarily change in shrub cover for this application, are more reliably quantified and mapped.

Besides achieving high effectiveness when producing shrub cover change estimates, other important considerations are cost efficiency and transferability of technology. We have emphasized and tested low-cost remote sensing approaches, since resource management and conservation biologist entities typically have limited budgets for environmental monitoring. As with many technologies, costs associated with remote sensing and image processing are "moving targets" and difficult to quantify in a meaningful manner. Also, trade-offs between achieving cost/efficiency and transferability, as well as accuracy and effectiveness are inevitable. These trade-offs are considered as we prescribe an operational shrub monitoring system in the sections that follow.

4.1 Aerial image capture

Based on our shrub classification and image change analyses reported above and in other studies [9, 15, 17], we continue to advocate for the use of aerial multi-temporal/multi-spectral imagery captured with small, piloted fixed-wing aircraft and digital single lens reflex (DSLR) cameras for monitoring vegetation change over extents on order 1 to 250 km^2. Small fixed-wing aircraft are relatively inexpensive and flexible platforms for aerial imaging. Relatively inexpensive moderate format DSLR cameras with 25 to 100 megapixel frame arrays can be modified to achieve single or multiple visible (red, green, and blue) and NIR waveband sensitivity, with two to four cameras integrated into aerial camera mounts and synchronized with location-based (GNSS) triggering systems. This means that very high spatial resolution, visible and near-infrared image frames covering large areal extents are collected, minimizing the number of images that are captured, stored and processed. Based on analyses summarized above and in Uyeda et al. [9] and Warkentin et al. [10], aerial images having a spatial resolution of c. 0.15 m.

Mission planning and image capture should be conducted consistently over space and time, in a manner that supports subsequent image pre-processing, classification and change analysis. Given the requirement to correct images for relief displacement with the aid of a DSM, a high amount (e.g., 70–80%) of along- and across-track image overlap (i.e., coverage redundancy per frame) must be planned and executed. Flight lines should be oriented in a direction that best maintains nearly constant heights above terrain. Pilots should attempt to maintain a nearly constant height above ground level. Images should be collected when herbaceous vegetation is senescent and shrubs have mostly green leaves, and when terrain shadowing is minimized. This would be in mid-summer (late July or early August) for the SCI study area. Image capture near mid-day will also minimize shadow effects. Data on camera station positions (x-y-z in the sky) and date/time of acquisition for each image frame should be archived with the original mission plan. Whether fixed or automatic aperture and shutter speed settings are utilized, the same radiometric control setting should be used for follow-on missions.

Subsequent imaging missions should follow an approach called repeat station imaging (RSI) [18, 19], where the same flight lines are re-flown with GNSS-supported pilot navigation systems, and images are captured at the same station positions based on GNSS-based camera triggering systems. The above prescriptions are idealized, and ability to follow optimal mission plans may be challenging. For instance, at SCI military operations, fuel and maintenance logistics, and weather (particularly marine layer) conditions can interfere with conducting imaging missions on a given day.

We also recommend that ultra-high spatial resolution (0.06–0.08 m) image frames be collected for selected sample areas (i.e., not wall-to-wall, overlapping coverage) within the full monitoring domain, to be used for reference data generation purposes. Selection of imaging locations should be based on representativeness of shrub cover distributions and evidence of more dynamic shrub change from the 2010–2018 retrospective analysis summarized above. The same imaging locations should be repeated throughout thc monitoring program. Classification training, accuracy assessment and shrub change data will be generated through visual interpretation of these ultra-high spatial resolution images.

To provide context, we estimate a cost of USD 10,000 (~USD 70 per km^2) for providing imagery required to monitor shrub cover for each mapping period for the entirety of San Clemente Island. This relatively low cost is possible with the use of imaging systems based on off-the-shelf digital single lens reflex cameras mounted on small single-engine aircraft. The cost is based on capture of highly overlapping visible and NIR imagcry with GSD = 0.15 m for orthoimagcry production, and sporadic coverage at GSD = 0.08 m for classification training and accuracy assessment purposes.

4.2 Image pre-processing

Preparing aerial multi-spectral images for subsequent shrub cover and change mapping (i.e., pre-processing) primarily involves geometric processing to co-align multi-temporal images, correct geometric distortions, geographically reference pixels to map coordinates, and stitch together individual frames to create an orthoimage mosaic. The other type of pre-processing, radiometric processing, which involves manipulation of the image brightness values associated with pixels, should not be required since the primary information processing requirement is image classification, which is conducted separately for each date of imagery.

The first steps of geometric pre-processing should be image registration, for both multi-spectral images captured by different cameras on the same date, and for time-sequential, RSI pairs. Frame-based registration is achieved through image warping transformations where one waveband (e.g., NIR) image is registered to an image from another waveband (or RGB Bayer array) image, or where a recently captured image frame is registered to the corresponding frame for an earlier captured image of an RSI pair. The warping transformations are calibrated or controlled by image match points, image features automatically detected within and co-associated between corresponding image frames. Software routines are available for automatically identifying corresponding image pairs, finding and reconciling match points, transforming a subject image to match a reference image, and outputting a registered image pair. We recommend that multiple camera images from a single date be simultaneously co-registered to their RSI pairs. For example, the two (RGB and NIR) image frames from a recent collection, should be registered to the corresponding two image frames from an earlier date.

For the start of the monitoring program, registered RGB and NIR frames of the original (time 1) multi-spectral images should be input to SfM photogrammetric processing to generate a DSM and a georeferenced orthomosaic, using one of several commercial software packages. A DSM is first produced by automatically finding match points (similar to the image registration process) between overlapping portions of image frames. The magnitude of parallax (i.e., relative relief displacement) between image frames for each match point is proportional to the terrain height of the scene relative to a height datum. This yields a 3-D point cloud of terrain heights. Very accurate and precise x-y-z positions for ground control points throughout a scene are needed to transform the point cloud into absolute positional coordinates. Such control points are generated through ground surveying with real-time kinematic GNSS units, where one or more field personnel find, survey and take multiple ground photos of discrete ground control features. The photos enable image analysts to co-locate ground control features on the aerial image frames, such that corresponding image (pixel) coordinates are related to ground coordinates.

The 3-D point cloud is then converted into a raster DSM through a surfacing process, with the DSM providing the basis for orthorectification of the registered aerial image frames. The relative geometry of overlapping image frames represented by the match points is exploited to estimate geometric parameters of the digital cameras (i.e., interior orientation) and the camera station positions in space and view geometry to the ground surface (i.e., exterior orientation). This interior and exterior orientation information, the DSM and ground control survey data enable each image frame to be rectified and georeferenced. The interior portions of corrected (rectified), overlapping image frames are then stitched together to create a single seamless orthomosaic. An important radiometric processing step that follows, is balancing of spatially-varying image brightness near the frame seams of the mosaiced images, which results from the unequal distribution of reflected light measured by digital cameras viewing the ground surface from different directions.

The cost associated with SfM processing of VHR image data sets to generate ortho-imagery and DSMs is a moving target that should continue to drop as the software evolves and analysts gain greater proficiency in utilizing such software. We estimate a cost for SfM processing to be similar image acquisition cost, or approximately USD 8000 for all of SCI. This does not include costs associated with requisite GNSS field surveys of control points, which could be conducted by the user's organization, likely on a one-time (first monitoring pass) basis.

4.3 Canopy height data generation

The importance of incorporating canopy height information from nDSMs when classifying shrub, herbaceous and bare cover types is documented above. So was the fact that nDSMs may be derived from LiDAR, SfM applied to overlapping aerial imagery, or a combination of the two. If affordable for a shrub monitoring program, we recommend conducting a one-time, high density LiDAR survey at the start of the program, from which a DEM and an nDSM are derived from the resultant point cloud. The LiDAR-derived DEM should continue to be beneficial over the life of the monitoring program for accurately representing ground surface elevations, which will generally not change over time.

DSMs produced in subsequent years from SfM orthoimage processing of aerial multi-spectral images will represent the most recent distribution of canopy heights associated with shrub growth or decline, with the LiDAR DEM providing the ground-based normalization to yield a new nDSM for each subsequent imaging date. Classifying ground from vegetation canopy points from a SfM-derived DSM is challenging, which is why the LiDAR-derived DEM should be worth the initial investment. The key for successful LiDAR DEM-SfM DSM integration when generating an nDSM is that both data sets are adequately ground-controlled and based on similar map projections and datums, so that they may be co-registered in an accurate manner.

4.4 Image classification

VHR aerial multi-spectral orthoimage and nDSM data sets for each year that is selected for shrub monitoring will be input to an image classification routine to generate maps representing shrub and other growth form and land cover classes. Classifying shrub species can also be attempted, particularly for the baseline (first) mapping effort. Some form of multi-pixel image classification approach is appropriate, such as OBIA or convolutional neural networks (described below), given the fine image spatial resolution relative to the size of individual or patches of shrubs. This means that contiguous groupings or segments of pixels are classified rather than classifying each pixel independently. We recommend that several, independent spectral indices (e.g., NDVI, Visual Atmospheric Resistance Index (VARI), and Normalized Difference Blue-Red Index (NDBR)) be incorporated as input to the classification routines, in addition to multi-spectral radiance and nDSM data.

We demonstrate in the two study components above that the OBIA approach using image segmentation and a ruleset/expert classifier is effective at mapping vegetation growth form types, and to some degree, shrub species, when applied to VHR aerial orthoimagery. Trimble eCognition is the most commonly available and utilized commercial software for OBIA processing. Its segmentation and classification routines are mostly expert-based, meaning that input and parameter choices are primarily selected based on experience and trial and error. "Process trees" developed with this software are readily modifiable sets of instructions that are applicable to sub-regions having differing vegetation composition and/or terrain and soil background characteristics, and readily transferable to other users (including less experienced ones). An advantage of the OBIA approach is that training data requirements and therefore field data collection is minimal.

Machine learning classifiers applied following image segmentation, or based on convolutional deep learning routines, have much potential for generating reliable growth form maps, but require extensive training data. We have yet to explore this potential, but doing so would entail generation of substantial training data through a combination of field sampling and interpretation of even finer spatial resolution imagery than the 0.15 m orthoimagery analyzed in this study.

4.5 Shrub cover and change mapping

The output products from image classification are growth form and land cover maps, which provide the basis for estimating shrub cover percentages on a regular grid. A growth form map is categorically aggregated and recoded to shrub and non-shrub classes. A regular grid of square polygons is overlaid on the shrub vs. non-shrub map, and the number of shrub pixels within each grid cell is tabulated and divided by the total number of pixels in the grid cell to yield a percent shrub cover estimate per cell. Based on Warkentin et al. [10] a grid cell size of 20 m appears to be appropriate, since shrub cover estimates were only slightly less accurate than with coarser 40 m grid, and the smaller grid size makes it more sensitive to relative changes in cover. Hamada et al. [20] tested 25 and 50 m grid cells for monitoring growth form cover and recommended a 25 m monitoring grid, which is similar to the 20 m size and would also be an appropriate alternative grid size.

Once the image capture, processing, and shrub cover mapping approach is replicated for subsequent years of the monitoring program, the resultant gridded shrub cover map is incorporated with the one from the most recent prior mapping effort to generate a shrub change map. This is done by subtracting percent shrub values of the older map from the current one on a grid cell basis to yield an absolute shrub cover change map. A relative shrub cover change map should also be produced by dividing the absolute change percentage by the earlier shrub cover percentage.

4.6 Analysis of shrub cover change

Maps of shrub cover percentage change should be displayed and analyzed in a manner that appropriately represents the accuracy and precision of the cover estimates and allows for visualization of shrub change patterns relative to vegetation communities, topography and disturbance histories. Interval classes for portraying colors on shrub cover change maps should have percentage change ranges that are in line with the degree of error and uncertainty associated with the estimates. For example, while grid cell changes may be estimated at decimal percentage levels, if the RMSE or MAE of shrub change is 5–10%, representing change intervals at ±10% may be appropriate. The change maps should be overlaid and visualized with vegetation community association maps and digital elevation maps to better understand biophysical relationships with shrub recovery and decline. Similarly, shrub cover change statistics should be generated, reported and assessed according to vegetation association types (zonal analysis) and units (focal analysis). It may be useful and appropriate to generate spatial pattern metrics from the shrub change maps, perhaps to highlight spatial hotspots of shrub cover increase and decline.

5. Conclusion

We explored the potential of using extant aerial orthoimagery time series to monitor shrub cover change within San Clemente Island, an island with a history of heavy human disturbance and a recovering semi-arid shrubland that is now dominated by herbaceous vegetation cover. We analyzed apparent shrub cover change using a set of annual multi-temporal V/NIR orthoimages for the period 2010 to 2018. These readily available orthoimage data sets consist of images captured during different seasons and spatial resolutions, making it challenging to monitor subtle changes in shrub cover in a reliable manner. We assessed how well nDSMs derived from highly overlapping VHR image sets represent canopy heights for image classification of shrub cover. Though the height accuracy and structural representation of the SfM-derived nDSMs were found to be inferior to a LiDAR-derived nDSM, they are much less expensive to generate and sufficiently reliable to provide an important input feature for image classification of shrub and non-shrub categories. The key design factor for an end-to-end aerial imaging system for monitoring shrub cover change is replicating image capture and processing parameters as much as possible. Such replication includes seasonality, sensor and sensor settings and viewing perspective for image capture, frame-based image registration, and image classification and areal unit for quantifying cover fractions. Our follow-on research plans will include testing a bi-temporal shrub cover change analysis, using the 2017 aerial orthoimage data set as a baseline for implementing a repeat station imaging approach to shrub monitoring.

Acknowledgements

This research was funded under the Cooperative Agreement Award (W9126G-15-2-0034) issued by the U.S. Army Corps of Engineers: Vegetation Mapping at NALF San Clemente Island, Naval Base Coronado, California, and Cooperative Agreement Award (W9126-18-2-0068): Shrub cover monitoring and sensitivity analyses of vegetation community maps for San Clemente Island, CA. Image interpretation assistance was provided by Hailee McOmber. Natural resource personnel of San Clemente Island assisted with field work and the accuracy assessments. Richard McCreight of NEOS, LTD. provided ultra-high-spatial-resolution aerial imagery.

Author details

Douglas Stow[1*], Kelsey Warkentin[1], Kellie Uyeda[1], John O'Leary[1], Julie Lambert[2], Andrew Loerch[1], Hsiao-chien Shih[1], Lloyd Coulter[1], Leslie Bolick[3] and Kimberly O'Connor[4]

1 Department of Geography, San Diego State University, San Diego, CA, USA

2 Soil Ecology and Restoration Group, San Diego State University, San Diego, CA, USA

3 US Pacific Fleet Natural and Cultural Resources Program, Command Southwest, Naval Base Coronado, Coronado, CA, USA

4 US Naval Facilities Engineering Command Southwest, Naval Base Coronado, Coronado, CA, USA

*Address all correspondence to: stow@sdsu.edu

References

[1] Corbin JD, D'Antonio CM. Competition between native perennial and exotic annual grasses; implications for an historical invasion. Ecology. 2004;**85**:1273-1283

[2] Tierra Data Inc. Integrated Natural Resources Management Plan. California: Naval Auxiliary Landing Field San Clemente Island; 2013

[3] Liu W, Wang J, Hu Y, Ma T, Otgonbayar M, Li C, et al. Mapping shrub biomass at 10 m resolution by integrating field measurements, unmanned aerial vehicles, and multi-source satellite observations. Remote Sensing. 2024;**16**:3095. DOI: 10.3390/rs16163095

[4] Hamada Y, Stow DA, Roberts DA, Franklin J, Kyriakidis PC. Assessing and monitoring semi-arid shrublands using object-based analysis and multiple endmember spectral mixture analysis. Environmental Monitoring and Assessment. 2012;**185**:3173-3190

[5] Lippitt CL, Stow DA, Roberts DA, Coulter L. Multidate MESMA for monitoring vegetation growth forms in southern California shrublands. International Journal of Remote Sensing. 2017;**39**(3):655-683

[6] Hamada Y, Stow DA, Franklin J. Quantifying biological integrity of California sage scrub communities using plant life-form cover. Journal of Mediterranean Ecology. 2009;**10**:19-32

[7] Lippitt CL, Stow DA, O'Leary JF, Franklin J. Influence of short-interval fire occurrence on post-fire recovery of fire-prone shrublands in California, USA. International Journal of Wildland Fire. 2013;**22**(2):184-193

[8] Maxwell AE, Warner TA, Vanderbill BC, Ramezan CA. Land cover classification and feature extraction from National Agriculture Imagery Program (NAIP) orthoimagery: A review. Photogrammetric Engineering & Remote Sensing. 2017;**83**(11):737-747

[9] Uyeda KA, Warkentin KK, Stow DA, O'Leary JF, Snavely RA, Lambert J, et al. Vegetation mapping using hierarchical object-based image analysis applied to aerial imagery and LiDAR data. Applied Vegetation Science. 2019;**23**(1):80-93. DOI: 10.1111/avsc.12467

[10] Warkentin K, Stow D, Uyeda K, O'Leary J, Lambert J, Loerch A, et al. Shrub fractional cover estimation and mapping of a recovering island shrubland based on airborne multispectral imagery and Lidar data. Remote Sensing. 2020;**12**:3608-3623. DOI: 10.3390/rs12213608

[11] Snavely RA, Uyeda KA, Stow DA, O'Leary JF, Lambert J. Mapping vegetation community types in a highly disturbed landscape: Integrating hierarchical object-based image analysis with LiDAR-derived canopy height data. International Journal of Remote Sensing. 2019;**40**(11):4383-4400

[12] Pérez-Luque AJ, Ramos-Font ME, Tognetti Barbieri MJ, Tarragona Pérez C, Calvo Renta G, Robles Cruz AB. Vegetation cover estimation in semi-arid shrublands after prescribed burning: Field-ground and drone image comparison. Drones. 2022;**6**:370. DOI: 10.3390/drones6110370

[13] Williams K, Hobbs R, Hambur S. Invasion of an annual grassland in Northern California by *Baccharis*

pilularis ssp. *consanguinea*. Oecologia. 1987;**72**:461-465

[14] Rose MB, Mills M, Franklin J, Larios L. Mapping fractional vegetation cover using unoccupied aerial vehicle imagery to guide conservation of a rare riparian shrub ecosystem in Southern California. Remote Sensing. 2023;**15**:5113. DOI: 10.3390/rs15215113

[15] Stow D, Hamada Y, Coulter L, Angeulova Z. Monitoring shrubland habitat changes through object-based change identification with airborne multispectral imagery. Remote Sensing of Environment. 2008;**112**:1051-1061

[16] Tarnavsky E, Stow D, Coulter L, Hope A. Spatial and radiometric fidelity of airborne multispectral imagery in the context of land-cover change analyses. GIScience & Remote Sensing. 2004;**41**:62-80

[17] Kerr A, Stow D. Optimizing radiometric fidelity to enhance aerial image change detection utilizing digital single lens reflex (DSLR) cameras. Photogrammetric Engineering and Remote Sensing. 2018;**84**:149-158

[18] Coulter L, Stow D, Baer S. A frame center matching technique for precise registration of multitemporal airborne frame imagery: Methods and software approaches. IEEE Transactions of Geoscience and Remote Sensing. 2003;**4**:2436-2444

[19] Stow D, Coulter L, MacDonald G, Lippitt CD. Evaluation of geometric capture and processing elements in the context of a Repeat Station Imaging approach to registration and change detection. Photogrammetric Engineering and Remote Sensing. 2016;**82**:775-788

[20] Hamada Y, Stow DA, Roberts DA. Estimating life-form cover fractions in California sage scrub communities using multispectral remote sensing. Remote Sensing of Environment. 2011;**115**:3056-3068

Chapter 7

Spatio-Temporal Mapping and Prediction of Red Tides (Harmful Algal Blooms) in the Coastal Areas Using Modis Derived Products

Shanmugam Jayalakshmi, K. Priyalatha, E. Bharathi and L. Malavika

Abstract

Harmful Algal Blooms (HABs) occur when toxin-producing algae grow excessively in a water body. It causes harmful effects on marine animals and people who consume seafoods. This study aims to monitor and map algal species and correlate their occurrence with the prevalent changes in the oceanic environment. Ramanathapuram, Tuticorin and 21 coastal island areas in South Tamil Nadu and the Gulf of Mannar have been creamed off their richness of diversity recently in the wake of the HAB wave. The work was carried out with the help of Aqua-MODIS level 2 products, including chlorophyll-a concentration (CC), sea surface temperature (SST), diffuse attenuation coefficient at 490 nm (KD at 490 nm), and fluorescence line height (FLH). Spatio-temporal mapping and prediction of bloom occurrences for the consecutive 5 years (2018–2022) were performed. The prediction model was created in Python using the K-nearest neighbor classifier, and training data were used to train the model. It was validated using the Cyanokhoj website, a product of Google Earth Engine. The results obtained from the model and Cyanokhoj were spatially correlated. The overall accuracy of the prediction model obtained was 81.23%.

Keywords: HAB, CC, SST, FLH, KD

1. Introduction

The correct circumstances can enable algae to grow out of control, for some of these blooms to release poisons that can kill fish, animals, and alien species, as well as possibly infect humans and potentially result in their demise [1]. Numerous algae species thrive when wind and water currents are favorable, according to studies. HABs (Red tides) may also be related to overfeeding in other situations. This happens when nutrients (mostly phosphorus and nitrogen) from lawns, farms, and other sources flow into bays,

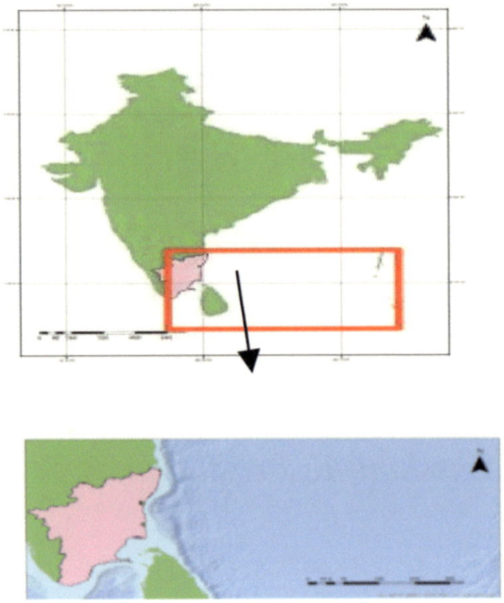

Figure 1.
Location of the study area.

rivers, and the ocean, and build up at a rate that overfeeds the algae naturally exist in the ecosystem [2]. A 30 km area, from close to Pamban to Keezhakarai, has recently seen an ecological dent in the coastal waters of Ramanathapuram and Tuticorin [3]. Water has suffered an ecological dent in a 30 km stretch from near Pamban to Keezhakarai [3].

The entire coast of Tamil Nadu and the Gulf of Mannar up to the exclusive economic zone (EEZ) are included in the study area (**Figure 1**). India's EEZ extends outside and next to its territorial waters, is 200 nautical miles from the International Date Line. The latitude and longitude of the study area are 15°N to 4.19°N, and 88°E to 77°E respectively. The Coromandel Coast region, which includes Tamil Nadu, stretches between Sri Lanka's west coast and India's southernmost tip.

2. Materials and methods

2.1 Dataset

Every day, images in 36 spectral bands and three spatial resolutions are produced by the MODIS sensor on board the Aqua spacecraft. For the 5 years from 2018 to 2022, monthly and daily data of CC, FLH, SST, and KD at 490 nm were acquired from NASA Ocean Color data and examined. Using SeaDAS and ArcGIS, HABs were mapped.

2.2 Methodology

Level 2 data of MODIS Aqua satellite of the National Aeronautics and Space Administration (NASA) for the period of 5 years (2018–2022) were downloaded and radiometric and atmospheric calibrations were done for the same. **Figure 2** depicts the process in detail. CC, SST, FLH, and KD490 nm were examined for the study period of 5 years. HABs were mapped based on the threshold values

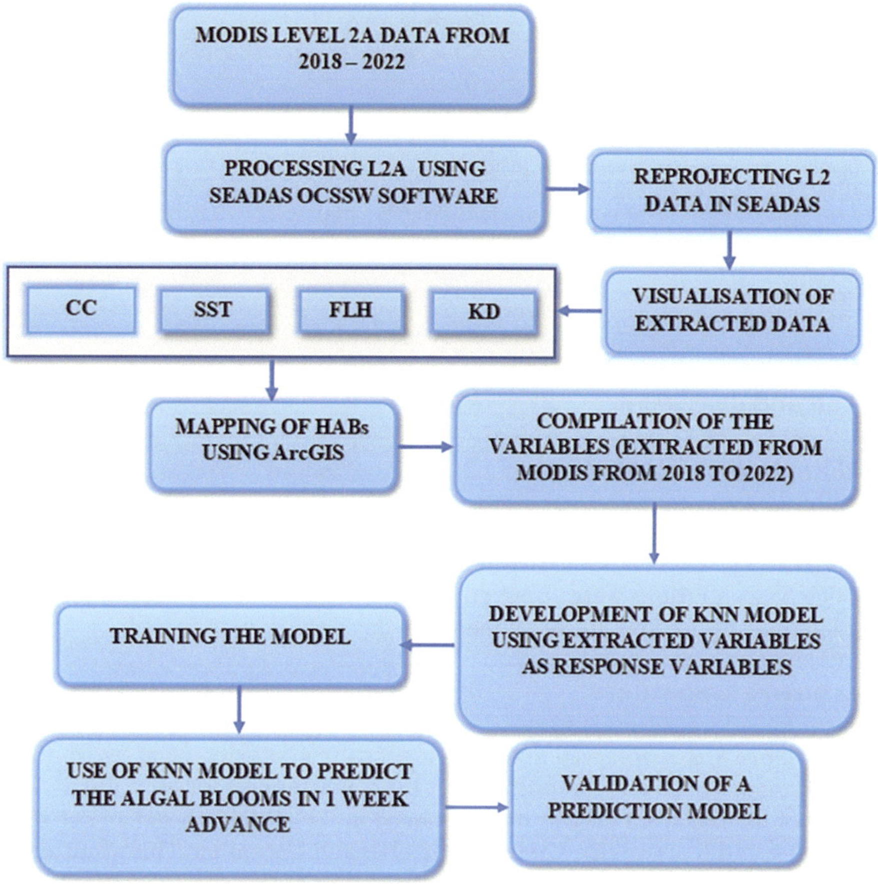

Figure 2.
Flowchart showing the general methodology.

of CC25–30°C [4], SST > =2.5 mg/m^3 [5], FLH > =0.0295 mW cm^{-2} m^{-1} sr^{-1} [6], and KD at 490 nm > =0.1236 [7] for a period of 2018–2021 using SeaDAS and ArcGIS software. To maintain consistency, the missing data values were set to 0.

Using the threshold values, the rasters were categorized and made into binary images. The binary pictures of CC, SST, FLH, and KD490 nm were used simultaneously after reclassification, and the pixels with the same value in all four rasters were found. The K-nearest neighbors classification technique was used to determine the algal bloom in advance of 1 week. The class with the highest nearest neighbors was given the data point assignment. Eighty percent of the original data has been used as training data and the remaining 20% as test data. The model was validated using the Cyanokhoj website, which uses sentinel photos to track the algal bloom [8]. The cell density of more than 2,000,000/ml is considered a dense bloom. Spatial auto-correlation was used to validate the prediction model's output. The tool determines the p-value, z-factor, and Moran I value. The Moran Index calculates the datasets' total correlation. The p-value represented the likelihood that the spatial pattern seen was the result of a random process. Moran I equal to +1 denotes positive auto-correlation, whereas the z-factor implies statistical significance, and vice versa (**Table 1**).

S. No.	Parameters	Threshold values	Author
1	Chlorophyll-a concentration	25–30°C	Beiras [4]
2	Sea surface temperature	$\geq$2.5 mg/m^3	Sudheeshan et al. [5]
3	Fluorescence line height	$\geq$0.0295 mW cm^{-2} µm^{-1} sr^{-1}	Coronado-Franco et al. [9]
4	KD at 490 nm	$\geq$0.1236	Sita Karki et al. [7]

Table 1.
Parameters and threshold values.

3. Result and discussion

3.1 Chlorophyll-a concentration

Less CC concentration has observed from **Figure 3** in the month of April and May in all the 5 years. This is because of the inverse relationship between SST and CC [10]. High concentrations were observed in the months of June to December in all the years. This may be because of low temperature during monsoon period.

3.2 Sea surface temperature

From **Figure 4**, it is observed that the temperature gradually increases from January through March and diminishes over time, reaching its lowest point in August as a result of breezes from the south-west monsoon. The pattern began to rise sharply in September and was sustained between October and December. This result also correlates with the CC concentration shown in **Figure 3**.

3.3 Fluorescence line height

Figure 5 displayed the FLH graph for a 5-year period (2018–2022). FLH is inversely related to SST, whereas it is directly related to CC. **Figure 5** is also showing less value in January to March and increased values in the period August. From **Figure 3** it is also observed that decreased CC during January to March and increased CC during August.

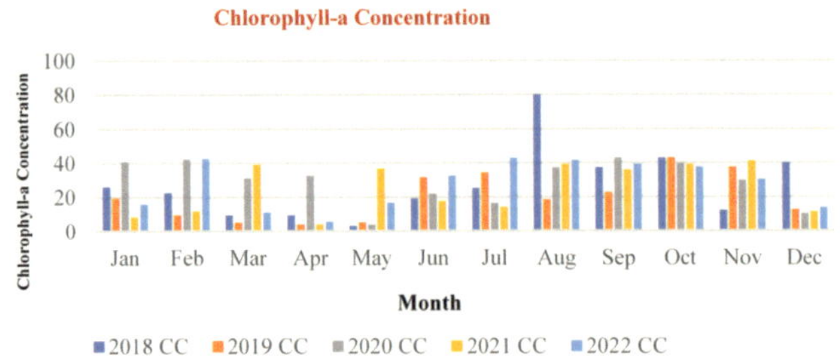

Figure 3.
Graph of mean CC from 2018 to 2022.

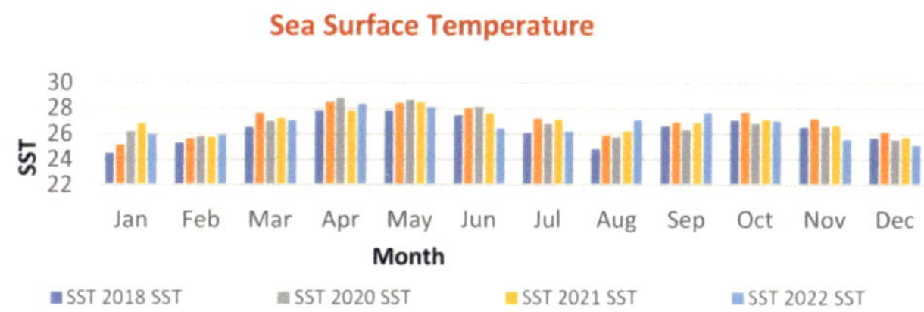

Figure 4.
Graph of mean SST from 2018 to 2022.

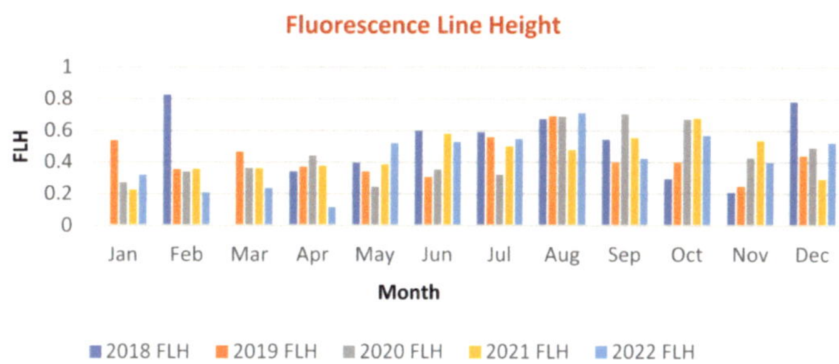

Figure 5.
Graph of mean FLH from 2018 to 2022.

3.4 Diffuse attenuation coefficient

From **Figure 6** it is observed that the KD has been dynamic for each month for all five consecutive years and does not follow a consistent pattern. This dynamicity is primarily due to phenomena such as high human activity, rivers, wind, wave-driven, and sediment re-suspension [11].

3.5 Threshold mapping of HABs

Figures 7(a-l)–10(a-l) illustrate the algal bloom from 2018 to 2021. Algal bloom has been observed in all the years except January and December, this may be due to the east coast's low salinity and the fact that algae grow best in freshwater.

Eddies and cyclones can uplift or transport nutrients from bottom layers to the surface, which might cause frequently occurring HAB species, like *Noctiluca scintillans* [12], to form blooms during cyclone-prone seasons [13–17]. These physical

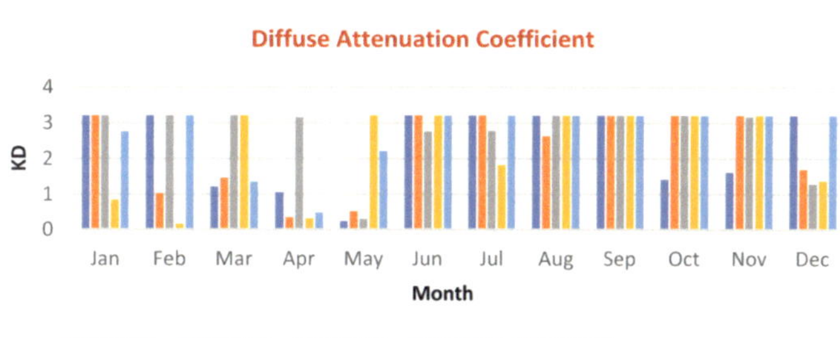

Figure 6.
Graph of mean KD490 nm between 2018 and 2022.

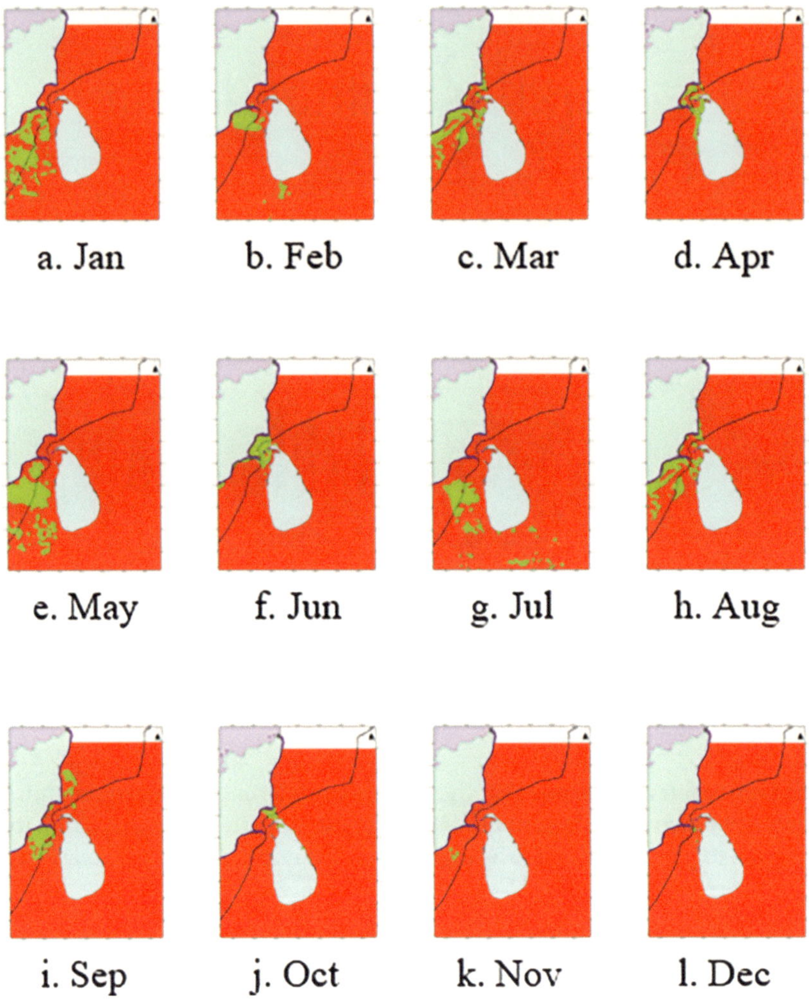

Figure 7.
Mapping of algal bloom for 2018.

mechanisms can also contribute to bloom formation in the Bay of Bengal. The period from March to May, known as the pre-monsoon season, had the greatest number of HABs occurrences.

Pre-monsoon weather was hot and muggy, which favored the algal bloom. Algal bloom incidences were found to be lower from June to September, the south-west monsoon season.

Even though there was a coastal up-welling because of the southerly winds, the up-welling was restrained by stratification, which prevents nutrients from deeper layers from reaching the top and reduces the amount of bloom occurrences during this time. Due to wind blowing from the North to the East, which in turn promotes coastal up-welling and brings nutrients from deep water to the surface, causing an algal bloom, the severity was intense throughout the Post-Monsoon and North-East monsoon period from October to November.

On the other hand, more algal blooms approximately six were reported in January 2018. This was as a result of La Nina being at its peak from November 2017 to January

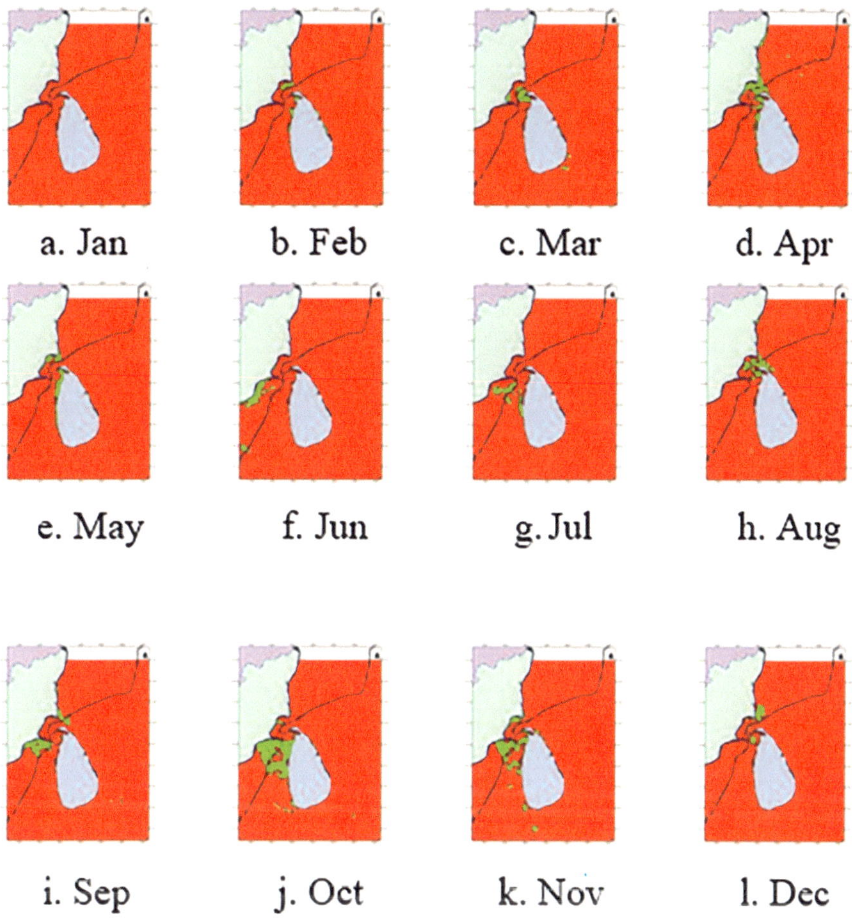

Figure 8.
Mapping of algal bloom for 2019.

2018, which coincided with the ENSO cold phase. La Nina increases the amount of nutrients reaching the coast, favoring algal bloom. Due to El Nino, fewer instances of algal bloom (five number approx.) were documented throughout October and November, which was unusual.

The warm phase of the ENSO cycle, known as El Nino, is when up-welling weakens or ceases entirely. From December 2020 to February 2021, there were around 10 fewer algal blooms due to a lack of nutrients from the deep. Due to the discharge of ballast water from cargo ships off the Sri Lankan coast, a deadly HAB wave occurred in the Gulf of Mannar in October 2021 in Pamban to Keezhakarai and in Mullaitivu [3].

3.6 Prediction model

The prediction model was developed using K-nearest neighbor classification model. The obtained Moran Index was 0.812345%. The computed p-value and z-score were 14.276974 and 0.000100, respectively. The test statistic value (p-value) of "0"

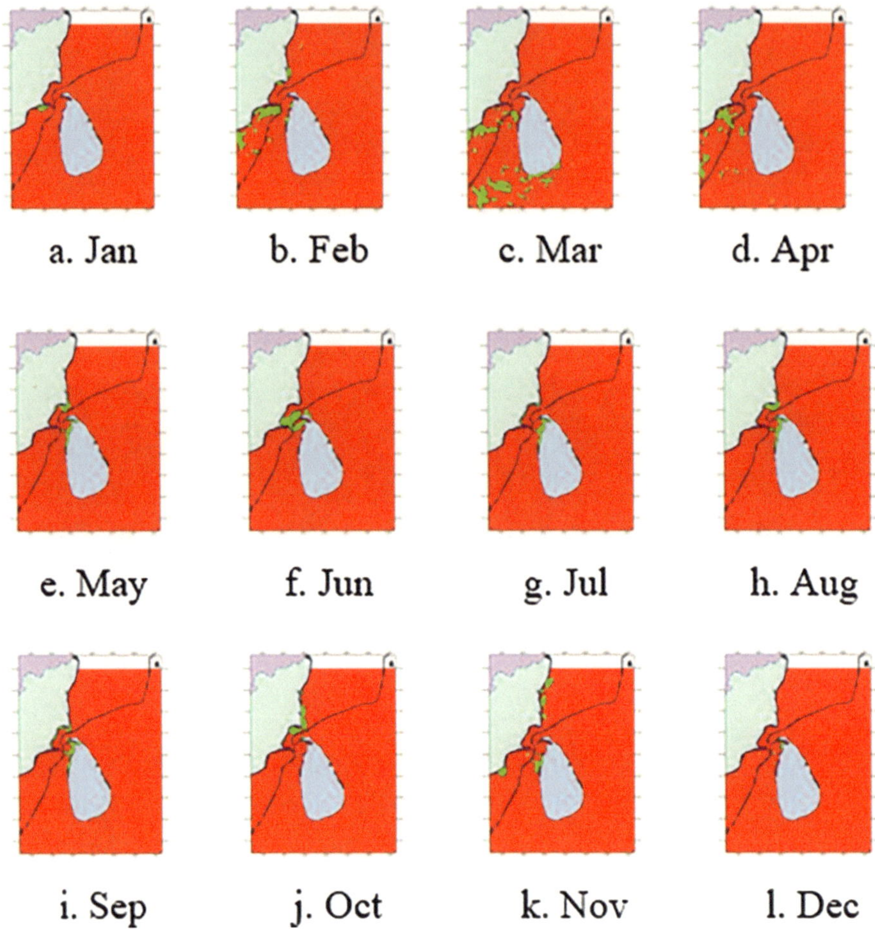

Figure 9.
Mapping of algal bloom for 2020.

denotes the presence of a fully positive correlation between the feature locations and the feature values.

There were two possible values for the z-score if the test statistic was 0.01. The possible values were that the z-score should either be larger than 2.56 or less than 2.58. The validation yielded a z-score of 14.276974, corresponding to a 99% confidence level. As a result, the prediction model's overall accuracy was 81.23%. **Figure 11** depicts the report produced by spatial auto-correlation. The results of the spatial auto-correlation were: 0.812345 for the Moran's Index, 14.276974 for the z-score, and 0.000100 for the p-value.

4. Conclusion

The spatio-temporal analysis for 4 years in a row on the coast of Tamil Nadu revealed that the blooming activity is predominating greater each year during the pre-monsoon and post-monsoon period. The accuracy of the test and training data were 84.54% and 81.13%, respectively. The prediction model was developed in a way that

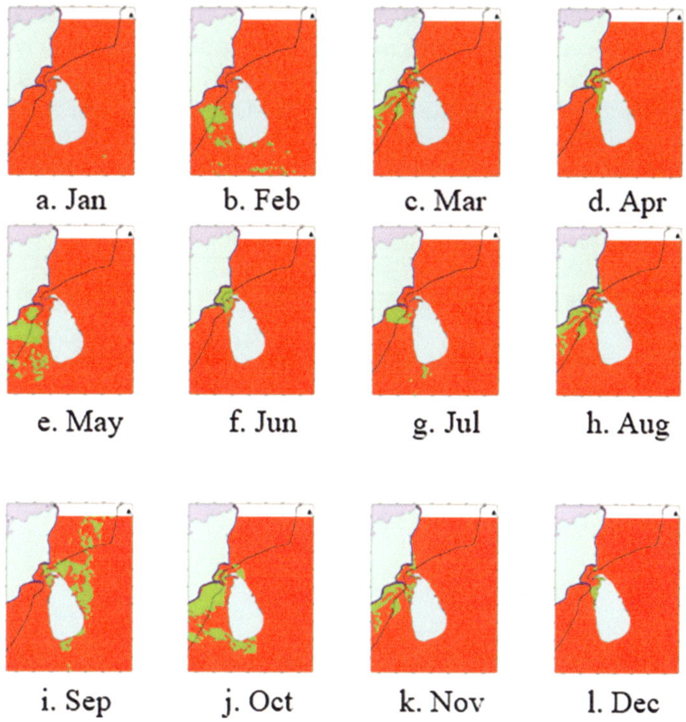

Figure 10.
Mapping of algal bloom for 2021.

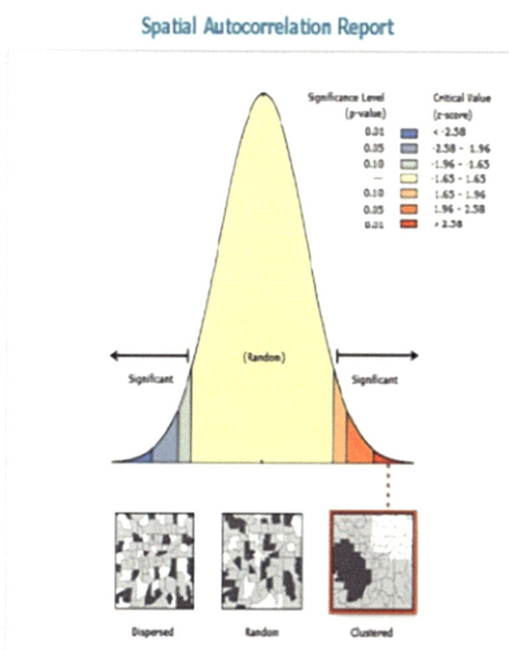

Figure 11.
Spatial auto-correlation report.

allows it to forecast 1 week in advance. As a result, the prediction model's overall accuracy is 81.23%. The accuracy of HAB's mapping can be increased when wind direction and wind speed are taken into account. With the aid of contemporary laboratory apparatus, the species present in the HABs can be identified. Higher resolution and less inconsistent satellite data might be taken into consideration for future research.

Author contributions

Priyalatha K, Bharathi E, and Malavika B. L (Students) performed all data simulations and analyses.

Jayalakshmi S (Professor) supervised and prepared the manuscript.

Conflict of interest

The authors declare no conflict of interest.

Author details

Shanmugam Jayalakshmi*, K. Priyalatha, E. Bharathi and L. Malavika
Institute of Remote Sensing, Department of Civil Engineering, College of Engineering Guindy, Anna University, Chennai, India

*Address all correspondence to: jayalakshmi@annauniv.edu

References

[1] National Oceanic and Atmospheric Administration Website, United States [Internet]. 2023. Available from: https://www.noaa.gov/what-is-harmful-algal-bloom [Accessed: March 15, 2023]

[2] United Nations Platform for Space-Based Information for Disaster Management and Emergency Response, Austria [Internet]. 2023. Available from: https://www.un-spider.org/category/disaster-type/harmful-algal-bloom [Accessed: March 15, 2023]

[3] Sundararaju V. Analysis of algal bloom in Gulf of Mannar and its effect on livelihood. Hindustan Times. 2019

[4] Beiras R. Nonpersistent inorganic pollution. In: Marine Pollution. 2018. pp. 31-39. DOI: 10.1016/B978-0-12-813736-9.00003-9

[5] Shaju SS, Akula RR, Jabir T. Characterization of light absorption coefficient of red *Noctiluca scintillans* bloom in the South Eastern Arabian Sea. Oceanologia. 2018;**60**(4):419-425. DOI: 10.1016/j.oceano.2017.12.002

[6] Coronado-Franco KV, Selvaraj JJ, Mancera Pineda JE. Algal blooms detection in Colombian Caribbean Sea using MODIS imagery. Marine Pollution Bulletin. 2018;**131**:791-798. DOI: 10.1016/j.marpolbul.2018.06.021

[7] Karki S, Sultan M, Elkadiri R, Elbayoumi T. Mapping and forecasting onsets of harmful algal blooms using MODIS data over coastal waters surrounding Charlotte County, Florida. Remote Sensing. 2018;**10**(10):1656. DOI: 10.3390/rs10101656

[8] Cyanokhoj Website, India [Internet]. 2023. Available from: https://chintanmaniyar.users.earthengine.app/view/cyanokhojindia [Accessed: May 5, 2023]

[9] Coronado-Franco K, Selvaraj J, Mancera J. Algal bloom detection using MODIS fluorescence data: A case study of the Colombian Caribbean Sea. Environmental Monitoring and Assessment. 2015;**187**(3):1-13. DOI: 10.1007/s10661-015-4327-6

[10] Hussein KA, Al Abdouli K, Ghebreyesus DT, Petchprayoon P, Al Hosani N, Sharif HO. Spatiotemporal variability of chlorophyll-a and sea surface temperature, and their relationship with bathymetry over the coasts of UAE. Remote Sensing. 2021;**13**(13):2447. DOI: 10.3390/rs13132447

[11] Yunus AP, Masago Y, Boulange J, Hijioka Y. Natural and anthropogenic forces on suspended sediment dynamics in Asian estuaries. The Science of the Total Environment. 2022;**836**:155569. DOI: 10.1016/j.scitotenv.2022.155569

[12] Raghu Prasad R, Jayaraman R. Preliminary studies on certain changes in the plankton and hydrological conditions associated with the swarming of Noctiluca. In: Proceedings of the Indian Academy of Sciences. Bangalore, India: Indian Academy of Sciences; 1954. pp. 49-57. DOI: 10.1007/BF03050400

[13] Gomes HR, Goes JI, Saino T. Influence of physical processes and freshwater discharge on the seasonality of phytoplankton regime in the Bay of Bengal. Continental Shelf Research. 2000;**20**(3):313-330. DOI: 10.1016/S0278-4343(99)00072-2

[14] Madhu NV, Maheswaran PA, Jyothibabu R, Sunil V, Revichandran C,

Balasubramanian T, et al. Enhanced biological production off Chennai triggered by October 1999 super cyclone (Orissa). Current Science. 2002;**82**(12):1472-1479. Available from: http://drs.nio.org/drs/handle/2264/287

[15] Naik RK, Hegde S, Anil AC. Dinoflagellate community structure from the stratified environment of the Bay of Bengal, with special emphasis on harmful algal bloom species. Environmental Monitoring and Assessment. 2011;**178**(1-4):15-30. DOI: 10.1007/s10661-010-1855-z

[16] Prasanna Kumar S, Nuncio M, Narvekar J, Kumar A, Sardesai S, de Souza SN, et al. Are eddies nature's trigger to enhance biological productivity in the Bay of Bengal? Geophysical Research Letters. 2004;**31**(7):L07309. DOI: 10.1029/2003GL019274

[17] Vinayachandran PN, Mathew S. Phytoplankton bloom in the Bay of Bengal during the northeast monsoon and its intensification by cyclones. Geophysical Research Letters. 2003;**30**(8):1572. DOI: 10.1029/2002GL016717